ANNALS *of* THE NEW YORK ACADEMY OF SCIENCES

T0188423

EDITOR-IN-CHIEF
Douglas Braaten

ASSOCIATE EDITOR
Rebecca E. Cooney

PROJECT MANAGER
Steven E. Bohall

EDITORIAL ADMINISTRATOR
Daniel J. Becker

Artwork and design by Ash Ayman Shairzay

The New York Academy of Sciences
7 World Trade Center
250 Greenwich Street, 40th Floor
New York, NY 10007-2157

annals@nyas.org
www.nyas.org/annals

**The New York
Academy of Sciences**

Published by Blackwell Publishing
On behalf of the New York Academy of Sciences

Boston, Massachusetts
2011

ANNALS *of* THE NEW YORK ACADEMY OF SCIENCES

VOLUME
1242

ISSUE

Annals Meeting Reports

TABLE OF CONTENTS

Ann. N.Y. Acad. Sci. ISSN 0077-8923

ANNALS OF THE NEW YORK ACADEMY OF SCIENCES
Issue: Annals *Meeting Reports*

Advances in bipolar disorder: selected sessions from the 2011 International Conference on Bipolar Disorder

David J. Kupfer,[1] Jules Angst,[2] Michael Berk,[3] Faith Dickerson,[4] Sophia Frangou,[5] Ellen Frank,[1] Benjamin I. Goldstein,[6] Allison Harvey,[7] Fouzia Laghrissi-Thode,[8] Marion Leboyer,[9] Michael J. Ostacher,[10] Etienne Sibille,[1] Stephen M. Strakowski,[11] Trisha Suppes,[10] Mauricio Tohen,[12] Robert H. Yolken,[13] L. Trevor Young,[6] and Carlos A. Zarate[14]

[1]University of Pittsburgh, Pittsburgh, Pennsylvania. [2]Zurich University Psychiatric Hospital, Zurich, Switzerland. [3]University of Melbourne, Melbourne, Australia. [4]Sheppard Pratt Health System, Baltimore, Maryland. [5]Institute of Psychiatry, Kings College London, London, United Kingdom. [6]University of Toronto Faculty of Medicine, Toronto, Ontario, Canada. [7]University of California at Berkeley, Berkeley, California. [8]F. Hoffmann-La Roche Ltd., Basel, Switzerland. [9]Hôpital Chenevier - Mondor, Universite Paris XII, Créteil, France. [10]Stanford University School of Medicine, Palo Alto, California. [11]University of Cincinnati Academic Health Center, Cincinnati, Ohio. [12]University of Texas Health Science Center, San Antonio, Texas. [13]The Johns Hopkins University School of Medicine, Baltimore, Maryland. [14]National Institute of Mental Health, Bethesda, Maryland

Address for correspondence: David J. Kupfer, MD, University of Pittsburgh, Western Psychiatric Institute and Clinic, 3811 O'Hara Street, Pittsburgh, PA 15213, kupferdj@upmc.edu

Recently, the 9th International Conference on Bipolar Disorder (ICBD) took place in Pittsburgh, PA, June 9–11, 2011. The conference focused on a number of important issues concerning the diagnosis of bipolar disorders across the life span, advances in neuroscience, treatment strategies for bipolar disorders, early intervention, and medical comorbidity. Several of these topics were discussed in four plenary sessions. This meeting report describes the major points of each of these sessions and included (1) strategies for moving biology forward; (2) bipolar disorder and the forthcoming new DSM-5 nomenclature; (3) management of bipolar disorders—both theory and intervention, with an emphasis on the medical comorbidities; and, (4) a review of several key task force reports commissioned by the International Society for Bipolar Disorder (ISBD).

Keywords: bipolar disorders; medical comorbidity; neuroscience; diagnosis

Session on strategies for moving biology forward

Overview

The aim of the session "Strategies for Moving Biology Forward," chaired by Marion Leboyer (Universite Paris XII), was to investigate new disease mechanisms such as immune–inflammatory markers and oxidative stress, while describing the utility of tools such as transcriptomics and multiomic profiling. Genomics has already pointed to diverse molecular pathways that confer risk to bipolar disorder, though not to the anticipated extent. Etienne Sibille described how he is developing methods to integrate results and potentially synergize the analysis of the Genome-wide Association Study (GWA) studies and transcriptomic studies. Coregulated RNA transcripts have been shown to identify coherent gene modules with shared functions, consequently, it has been hypothesized that the coregulated gene modules that are enriched in genes associated with GWA results may identify relevant biological pathways to explore the underlying mechanisms of bipolar disorder. Robert Yolken and Faith Dickerson have shown that acute mania was associated with evidence of immune activation by: (1) elevated levels of cytokines, such as TNF-α; (2) inflammatory macromolecules, such as C-reactive protein (CRP); (3) antibodies to infectious agents, such as retroviruses; (4) antibodies to food antigens, such as gliadin; and (5) antibodies to brain proteins, such as the NR-2 peptide of the NMDA receptor. In many cases, the

doi: 10.1111/j.1749-6632.2011.06336.x

levels of these biomarkers are elevated during acute mania and are decreased six months later. Ongoing research should lead to the identification of a biological "signature" of mania, offering potential tools both for diagnosis of acute mania and for prediction of illness course, and hope for new therapies. Medications directed at the modulation of the immune response may form the basis of a new therapeutic armamentarium for the prevention and treatment of this disorder. Trevor Young presented work in the field of mitochondrial dysfunction and energy metabolism in bipolar disorder. Data clearly show that oxidative damage is present in the postsynaptic membranes and also found in the periphery. Indeed, several studies report increased oxidative stress in serum and plasma of bipolar patients. From a therapeutic perspective, this pathway may be of great importance as several mood stabilizers have antioxidant properties. The conclusion of this session is that studying genetic, immunoinflammation, and mitochondrial dysfunction will be very helpful to better understand the pathophysiology of bipolar disorder.

Genome-wide association studies

Etienne Sibille (University of Pittsburgh) opened this session and described a series of fascinating studies. Ranscriptome (the set of all expressed genes in a tissue sample) and genome-wide association (GWA) studies have separately provided clues to mechanisms of neuropsychiatric disorders, although not to the anticipated extent. Transcriptome studies mostly focus on changes in gene expression in disease states (*altered expression*), but also provide unique opportunities for assessing the less-investigated changes in the coordinated function of multiple genes (*altered coexpression*). Moreover, results from these large-scale investigations of changes in gene function are poised to interact with GWA studies of structural changes in genes and regulatory regions (DNA variant). Methods for integrating these approaches are now being developed.

Gene arrays allow for the unbiased quantification of expression (mRNA transcript levels) for 10,000 to 20,000 genes simultaneously (see Fig. 1A). Since gene transcript levels represent the integrated output of many regulatory pathways, the study of all expressed genes provides an indirect snapshot of cellular function under diverse conditions. For instance, using postmortem brain samples, this "re-

verse engineering" approach has implicated mitochondrial dysfunction and immune/inflammation-related changes in bipolar subjects.[1–3] However, current studies are still very few, were performed in heterogeneous cohorts, and utilized early and rudimentary versions of gene arrays. Moreover, gene array studies are subjects to similar limitations as GWA studies, in that large number of genes are tested ($n = 20$–40,000) in few subjects ($n = 10$–100). Typically, results identify 1–10% of genes affected in the illness, are characterized by very high rates of false discovery, and often do not properly account for numerous clinical (e.g., drug exposure, subtypes, duration), demographic (age, sex, race), technical parameters (RNA integrity, brain pH, postmortem interval for brain collection), or other potential cosegregating unknown cohort specificities. Conditions of postmortem brain collection also preclude the reliable identification of acute state-dependent gene changes, but are appropriate for investigating stable long-term disease-related homeostatic adaptations.[4]

Transcriptomics

A notable derivative and less investigated aspect of transcriptome studies is the development of gene coexpression studies. Here, two genes are defined as *coexpressed* in a dataset if their patterns of expression are correlated across samples (see Fig. 1B). Coexpression has been shown to reflect shared function between these genes, and may arise through multiple biological pathways, including cellular coexpression and common regulatory pathways (e.g., hormone signaling, transcription factors).[5,6] Hence, coexpression links have been used to build gene networks, and to identify communities, or modules, of genes with shared functions (see Fig. 1B).[7,8] Notably, by incorporating multiple interactions among large number of genes, the study of gene coexpression networks provides one solution to tackle the complexity of biological changes occurring in complex polygenic disorders.[5] Indeed, the information content of a large-scale gene network could be compared to the operation of a symphonic orchestra of many hundreds of instruments, where the intrinsic balance (coexpression) between instruments (genes) provide harmonious (homeostatic) function.[9] Efforts from our research team has demonstrated that brain gene coexpression networks assemble into small-world and scale-free networks,[10] an efficient

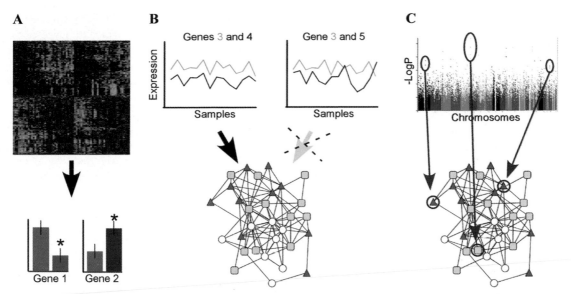

Figure 1. (A) Altered gene expression. Large-scale gene arrays simultaneously investigate expression levels of thousands of genes (top) and seek changes in gene transcript levels between conditions (bottom). (B) Gene coexpression. (top) Genes 3 and 4 display correlated patterns of expression across samples (X axis) between two conditions (gray and black lines). This relationship is quantified by Pearson correlation and provides a link in the gene coexpression network (bottom). Genes 3 and 5 are not coexpressed and do not contribute to the network. Nodes (genes) of different colors and shapes are associated with different cellular functions.[5,119] (C) Synergizing GWA and transcriptome studies. Genes associated with candidate DNA variants, as identified by GWA Manhattan plots (top), are overlaid onto gene coexpression networks (bottom). Gene coexpression networks that are enriched in GWA-associated genes are hypothesized to represent candidate biological systems for disease mechanisms.

network topology that is often observed in biological systems, but also in communication and social networks.[9] Interestingly, our initial study suggests that the structure of brain gene coexpression networks is largely resilient to changes occurring in neuropsychiatric disorders (major depression, bipolar depression, and schizophrenia), and instead points to a peripheral network localization of differentially-expressed genes.[10] This is in striking contrast to cancer- and other systemic illness-related networks, where affected genes tend to display a "disease/lethality–network centrality" relationship.[11] This contrast may relate to the difficulty in identifying silver-bullet types of drugs in neuropsychiatry, since affected genes may attend to control biological gene networks from the periphery, rather than through central gene hubs.[10]

Based on the assumptions that (1) hits on different components of a biological pathway may lead to the same disease phenotype, even in the absence of common changes across subjects, and (2) that DNA genetic polymorphisms and gene mRNA transcript levels represent independent and complementary measures of gene structure and function, we have

hypothesized that modules of coexpressed genes, which are enriched in genes associated with GWA-identified polymorphisms, may represent candidate biological pathways for recruitment in mechanisms of disease (see Fig. 1C). To address this question, our approach has been to identify conserved and robust gene modules based on multiple and extensive transcriptome studies in the human brain and to overlay onto those modules, genes that are located nearby DNA polymorphisms associated with neuropsychiatric disorders, as identified by GWA studies. Current efforts are addressing the difficulties in combining heterogeneous studies and the statistical and analytical challenges that emerge from integrating multiple large-scale approaches (transcriptome, coexpression network, and GWA).

In summary, the potential of transcriptome studies for unbiased novel discovery and for investigating basic pathological changes beyond the usual suspects is vast, but not yet realized in neuropsychiatry. Early results from transcriptome studies of altered gene expression in bipolar disorders and other major mental illnesses are promising, but results need to be replicated across multiple cohorts and research groups. More studies are needed using updated genetic information and technological platforms. Shared access to the raw data is also necessary, as the next generation of transcriptome studies will need to apply novel statistical methods for within-study parameter integration and for across-study meta-analyses, including permutation-based methods and accurate control of false discovery. Currently, the study of gene coexpression networks in neuropsychiatric disorders is still in its infancy, but the field will benefit from applying network methodologies developed for investigating other complex biological systems, including development, cancer, and brain functional activity. Finally, concepts and methods for integrating functional (transcriptome) and structural (DNA polymorphism GWA) studies of the molecular bases of complex neuropsychiatric disorders need to be developed to harness the potential of systematic large-scale molecular and genetic investigations of the brain.

Immune–inflammatory markers

Robert Yolken (Johns Hopkins University) and Faith Dickerson (Sheppard Pratt Health System) then described some of their new work. Mania is an abnormal mood state, and the defining characteristic of bipolar disorder in which the etiology is unknown. Immunological abnormalities have been identified which may contribute to the pathophysiology of mania as well as to bipolar disorder more broadly.[12–16] Such factors may help explain the marked fluctuations in mood symptoms, which are the hallmark of the disorder.

In a previous study, they examined the level of CRP, a nonspecific marker of inflammation in individuals with bipolar disorder.[17] CRP is a pentameric protein that is generated in the liver and secreted in the blood. The measurement of CRP in the blood provides a reliable marker of chronic inflammation caused by infectious and other inflammatory agents. We measured the level of CRP in $N = 122$ outpatients with bipolar disorder and $N = 165$ control individuals and evaluated the symptom severity of the bipolar disorder patients. Within the bipolar disorder sample, CRP was significantly associated with the Young Mania Rating Scale (YMRS)[18] score and in a multivariate analysis, CRP was the only independent predictor of YMRS score. The CRP levels of the $n = 41$ individuals with YMRS > 6 were significantly greater than the levels of the $n = 81$ individuals with YMRS $\leq$, 6. The CRP levels of the group with YMRS > 6 were also significantly greater than the levels of the control group while the CRP levels of the group with YMRS ≤ 6 did not differ from that of controls.

Based on the results of this cross sectional study, they undertook a longitudinal study of individuals hospitalized for symptoms of acute mania. Our aim was to measure inflammatory markers in acute mania and to determine changes over time in these markers and also to determine the correlation of markers with clinical outcome, whether or not persons were rehospitalized for a new illness episode in the six-month follow-up period. From blood samples we measured inflammatory markers including antibodies to intestinal antigens, antibodies to neuroreceptors, antibodies to endogenous retroviruses, and cytokines. The sample of $N = 60$ participants had a mean age of 35.4 (SD $= 12.9$) years and was 30% male and 67% Caucasian. The diagnoses of study participants were divided among bipolar disorder, most recent episode manic (55%), most recent episode mixed (34%), and schizoaffective disorder, current manic episode (11%). The mean YMRS score at the time of evaluation during the hospital stay was 18.7 (SD $= 8.6$) and the Positive and Negative Syndrome Scale total score[19] was 75.2 (SD $= 11.7$). For most study participants, blood samples were measured at three time points: the day of hospital admission from archived samples from the medical admission work-up ($n = 44$); the day of evaluation for the current study, on average 3–5 days following hospital admission ($n = 60$); and at a planned six month follow-up ($n = 39$).

Results of the longitudinal study show that patients hospitalized with mania had increased levels of immune markers that were lower six months later. Some of the markers appear to be mania specific. Some of the markers were associated with clinical outcome, whether or not patients were rehospitalized for a new illness episode during the follow-up period.

The results of these studies are consistent with a body of literature suggesting that acute episodes of mania are associated with evidence of immune activation. The pathways, which are suggested to be involved, are shown in Figure 2. The literature indicates an association between mania and elevated levels of cytokines including interferon γ,

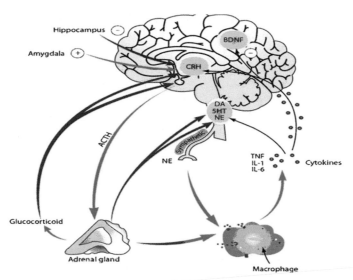

Figure 2. Molecular processes are affected by stress and depression. From Ref. 120, with permission from John Wiley and Sons.

interleukin 2 (IL-2), IL-6, TNF-α;[20] cell-mediated immunity, including the composition of lymphocytes and T cell receptors;[21] complement such as immune complexes and C1q,[16] and lipid levels such as cholesterol.[22] Studies indicate that many of these abnormalities normalize following resolution of acute mania and differ from patterns found in bipolar depression. Our data support the study of interventions addressed at lowering levels of markers of immune activation as a possible treatment for mania. This research may lead to improvements in the diagnosis and treatment of bipolar disorder.

Trevor Young (University of Toronto) presented the final report in this session in which he reviewed the evidence for mitochondrial dysfunction and oxidative stress in this illness. Earlier findings of maternal inheritance in bipolar disorder suggested the importance of mitochondria[23] and the field moved forward quickly with the finding of increased mitochondrial DNA deletions in postmortem brain.[24] It was then quite intriguing that several microarray studies on independent samples of brain tissue found remarkable decreases in mRNAs for components of the electron transport change,[25] which is one of the core functions of the electron transport chain. Their lab has followed up on these data and found compelling evidence of mitochondrial dysfunction in bipolar disorder. They reported decreased levels of one of the components of complex I of the electron transport chain and also the activity of this complex in prefrontal cortex obtained

postmortem from patients with bipolar disorder.[26] There was also increased protein carbonylation and 3-nitrotyrosine levels in the same samples. There was a significant inverse correlation between complex I activity and the oxidative metabolites in the same brain region. They next measured the brain's antioxidant system in the same samples and found that in bipolar disorder, as in schizophrenia and major depression, that the mean levels of glutathione were depleted in all diagnostic groups suggesting an impact of chronic psychiatric illness on this system.[27] Interestingly, the same findings appear to be present in another postmortem brain sample from subjects with bipolar disorder and the differences are particularly evident in the synaptosomal fractions (Andreazza *et al.*, unpublished data), which suggests that oxidative damage might be present in postsynaptic membranes and could affect function at this level which would be consistent with traditional sites of pathophysiology in mood disorders.

What is particular fascinating and might be relevant to move the field forward is that many of these findings are also found in blood samples from patients with bipolar disorder. Indeed, Konradi *et al.*[2] found that the expression of a number of electron transport chain mRNAs was altered in lymphocytes from patients with bipolar disorder. More recently, mitochondrial clustering has been demonstrated in lymphocytes and fibroblasts from patients with bipolar disorder.[28] There have been a number of findings of increased oxidative stress in serum and plasma markers and a meta-analysis has shown that several of these findings are evident across studies.[29] These data have been particularly important in the work of one of the International Society of Bipolar Disorders (ISBD) committees on the study of biomarkers (see below). Finally, this pathway might also be relevant for treatment of bipolar disorder since several mood stabilizers, lithium, and valproate, have antioxidant properties,[30] and at least several studies suggest that the antioxidant, N–acetylcysteine, may have some mood stabilizing properties.[31] The conclusion of this presentation is that, indeed, continued study of mitochondrial dysfunction may be very helpful to gain a more fulsome understanding of the pathophysiology of bipolar disorder.

Session on diagnosis

Overview

The session on diagnosis was chaired by Carlos Zarate (National Institute of Mental Health). The concept of bipolar disorder, initially known as manic-depressive insanity, has gone through considerable revision since it was first proposed.[32] Bipolar disorder consists of two major components, depression and hypo/mania, which vary greatly in severity, duration, and course. In addition, these components may present simultaneously or separately in time. Over the years, the boundaries of depression, mania, and bipolar disorder continue to shift. In recent years, attention has been given to the concept of bipolar spectrum disorders.[33] The DSM was introduced in part to clarify the ambiguities in psychiatric diagnosis. However, even with the advent of the DSM, there remains considerable uncertainty on diagnostic concepts of bipolar disorder.

The DSM is now going into its 5th edition, and the bipolar disorder subworkgroup in collaboration with the full Mood Disorder Workgroup for the DSM-5 has come up with proposed revisions addressing some of these diagnostic ambiguities. It is important to note that when there is a proposed revision that this is done by following a series of revision principles and utilizing generally accepted validators of a diagnostic entity. The principles of revision used in the workgroup were to optimize clinical utility, to make recommendations guided by research evidence, to maintain continuity with previous editions, and to set the stage for future developments in our understanding of the brain. For there to be a change in the criteria it is also necessary to elucidate the reason for change and present evidence in support of change. Jules Angst summarized work and research on evidence-based efforts to redefine bipolar disorders, Ellen Frank discussed revisions to the concept of mixed episodes in DSM-5, and Trisha Suppes covered criteria of hypomanic episodes.

The report by Angst uses a new specifier (increased energy/activity levels) to understand the associations among major depressive disorder, bipolar I, and bipolar II disorders. When using this specifier in the BRIDGE study (bipolar disorders: improving diagnosis, guidance, and education), what was found was that a significant number of individuals in a major depressive episode diagnosed with major depressive disorder in reality had a bipolar disorder diagnosis. When using the specifier criteria, the authors found that the distinction between BP-I versus BP-II disorders was much sharper. BP-I disorders had higher suicide attempt rates, whereas individuals with BP-II disorders had a greater association with anxiety disorders. Also, both bipolar I and II disorders, when compared with MDD, had higher rates of association with social phobia, substance use disorders, obsessive-compulsive disorder, suicide attempts, and attention disorder hyperactivity disorder. BP-II disorder compared to MDD had significantly more generalized anxiety disorder and panic.

Mixed episode, initially described as mixed states of manic-depressive insanity, has received many revisions since originally conceptualized.[32] Early on, it was evident that some patients with acute mania or hypomania simultaneously experience prominent depressive symptoms. Clinicians tried to capture this ambiguity that surrounds this condition by developing diagnostic criteria. A number of diagnostic criteria have been proposed over the years.[33] DSM-IV criteria specified that in order to meet criteria for a mixed episode that an individual was required to simultaneously fulfill criteria for both a manic and major depressive episode nearly every day for a period of at least one week. This was viewed as too broad and is often not seen. Instead clinicians were more likely to encounter individuals with a simultaneous admixture of depressive and manic symptoms that did not meet DSM-IV criteria for a mixed episode. Such clinical presentations were also associated with many of features of this more stringent form (i.e., more likely to be associated with suicidality, longer duration of illness, concomitant alcohol or sedative-hypnotic abuse, poorer outcome, and less adequately responsive to lithium). Thus, identifying the mixed state presentations across the spectrum would likely have major clinical implications. The paper by Frank *et al.*[59] proposes that mixed states be identified by *specifiers* that can be applied to episodes of either depression or mania/hypomania and that can be applied to individuals with a lifetime diagnosis of either major depressive disorder or bipolar disorder. It is believed that the use of specifiers would lead to earlier diagnosis and treatment.

Another challenge in the classification of bipolar disorder is how to define a hypomanic episode. The paper by Trisha Suppes and colleagues[34] addresses this important issue. Currently in DSM-IV-TR, the prototypical symptom of hypomania is elevated mood (and/or irritable mood). The DSM-5 bipolar subgroup proposes to take out one of the criterion B symptoms, "increase in goal-directed activity," and to place it with "elevated mood" in criterion A. Thus, the change in criterion A would be a distinct period of abnormally and persistently elevated, expansive, or irritable mood AND change in activity/energy levels, which should also be abnormally and persistently increased. This suggested change would lead to increase specificity without loss of sensitivity. The other issue addressed in the paper by Suppes is the number and duration of hypomanic/manic symptoms necessary to meet criteria for a hypomanic episode. Clinicians are often unsure how to proceed when diagnosing a hypomanic episode. This is not surprising as diagnosis of hypomanic episode is the most unreliable using DSM-IV criteria. There has been an abundance of studies looking at whether hypomanic symptoms should be present for two, three, four, or more days in order for the diagnosis to be made. The workgroup proposes to keep the four-day requirement, which would increase specificity with reasonable sensitivity.

The proposed DSM-IV changes are an important step to resolve some of the ongoing diagnostic ambiguities of bipolar disorder. Clearly, the proposed changes are in need of testing, which will occur under the field trials. The DSM-5 committee is open for comment and considering changes in diagnostic criteria for bipolar disorder in light of these proposals (www.dsm5.org).

Mood disorder workgroup reports

Jules Angst presented first. Conventionally, mood disorders are classified into depression, mania, and bipolar disorders. While these distinctions are very useful for international communication and treatment decisions, nature is more complex. The two components, depression and mania, vary greatly in severity and course, as do the subgroups of bipolar disorders (BP-I, BP-II, Md (mania with minor depressive disorders)).[35] As Cassano *et al.*[36] demonstrated in regard to patients with recurrent major depression, there is a linear increase in the number of manic symptoms as a function of depressive symptoms over the patient's lifetime. Bipolar disorders often manifest first as depression; we know for a fact that their diagnosis may often be delayed by up to ten years[37] and that depression in patients with bipolar or subthreshold bipolar disorder is less responsive to antidepressants.[38,39]

Depression has long dominated the field in terms of prevalence, treatment and, especially, in the estimates of global burden of disease. But if we assume a spectrum from depression via bipolar disorders to mania, it is important to investigate how many subjects suffering from major depressive disorders also manifest sub-threshold hypomania. In recent years, reanalysis of two large epidemiological studies (Early Developmental Stages of Psychopathology (EDSP) study in Munich, Germany and the National Comorbidity Study-R in the United States) found that as many as 40% of patients with DSM-IV major depressive disorder (MDD) met criteria for sub-threshold bipolarity; the latter was strongly predictive for BP-I disorder ten years later.[40,41]

The structured interviews (DIS) applied in those studies did not assess sub-threshold hypomania in any detail. This gap has been filled by the BRIDGE study, which made detailed assessments allowing an operational definition. The study comprised 5,635 patients with DSM-IV major depressive episodes (MDE) recruited in North Africa, Europe, and the Near and the Far East.[42] The study validated a new "specifier" (S) definition for hypomania/mania.[43] This adds increased activity/energy to DSM-IV criterion A (elated or irritable mood) and eliminates all the DSM exclusion criteria (e.g., hypomania under antidepressants) because they rule out patients who have a clinical profile of bipolarity. The other criteria for BP-I-S are identical to those of DSM-IV. BP-II-S, on the other hand, differs from DSM-IV BP-II by requiring a hypomanic episode duration of only one day or longer (instead of the 4+ days in DSM-IV) and the lack of any exclusion criteria. The validity of short hypomanic episodes of one and 2–3 days was demonstrated by the BRIDGE study.[42]

The consequences of redefinition by the specifier criterion are very considerable: not only does it identify many more patients with affective disorders as having bipolar disorders, it also distinguishes more clearly between the new bipolar subgroups and the remaining MDD in terms of validators and other clinical characteristics (comorbidity, nonresponse to antidepressants). The current DSM-IV criteria diagnosed 4,732 (84%) of the 5,635 patients with a MDE as having MDD, but the specifier criteria diagnosed only 2,988 (53%) (MDD-S). Forty-seven percent were diagnosed as suffering from bipolar disorders. The specifier concept turned out to be more valid than the DSM-IV classification, as illustrated by the frequent presence of a family history of mania, a progressively more recurrent course, more full remission between episodes, and other clinical characteristics for bipolar disorders.[42]

DSM-IV diagnosed 12.1% of the 5,635 MDE patients as having BP-I and 2.3% as having BP-II; specifier criteria diagnosed 23.9% as having BP-I-S and 23.1% as having BP-II-S. The distinction between BP-I versus BP-II disorders was much sharper when specifier diagnoses were applied. BP-I-S disorders were associated with higher suicide attempt rates, and BP-II-S disorders to a greater extent with anxiety disorders (generalized anxiety disorder (GAD), panic disorder (PD), and obsessive-compulsive disorder (OCD)). In comparison with MDD-S both BP-I-S and BP-II-S disorders differed in their higher rates of association with social phobia, OCD, binge eating, suicide attempts, substance use disorders, ADHD, and borderline personality disorder. Whereas, compared to MDD-S, BP-II-S disorder was significantly more associated with GAD and PD, there was no difference at all between BP-I-S disorder and MDD-S in this respect. The strong comorbidity of BP-II-S disorders with anxiety states is clearly of great clinical interest. Figure 3 illustrates the enormous differences between the varying concepts in the diagnosis of bipolar disorders. The DSM-IV definition gave the low-

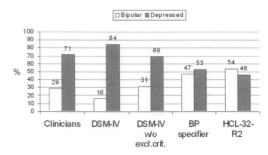

Figure 3. Diagnosis of bipolar versus depressive disorder by varying definitions in the Bridge Study of 5,635 patients with MDE.

est rates; these doubled with the omission of the exclusion criteria and tripled with application of the specifier criteria. Self-assessment using the Hypomania Checklist 32-R2 as a screening instrument yielded the highest rates. It is of particular interest that the clinicians' initial diagnosis identified almost twice as many patients as bipolar compared to the DSM-IV criteria.

In conclusion, the specifier definition of hypomania identifies with good validity many more patients with MDE as having bipolar I and bipolar II disorders and thus allows a much earlier diagnosis and hopefully a better treatment. The findings need replication by other studies. The DSM-5 committee is considering changes in diagnostic criteria for bipolar disorder in light of these findings (www.dsm5.org). Changes may allow a better characterization of major depression as either unipolar or bipolar and give consideration to dimensional issues.[44] This is particularly relevant to the definition of mixed states. While the results of the Bridge study and the application of the bipolarity specifier should not necessarily change the overall proportion of subjects from the general population who suffer from affective disorders, it may have an impact on the proportion of these affective patients who are recognized as belonging to the bipolar spectrum, as opposed to the unipolar spectrum. The impact of such change on treatment and outcomes should be monitored over the next decade.

Ellen Frank (University of Pittsburgh) was the second presenter in this session. In order to meet criteria for a mixed episode in the DSM-IV, an individual was required to *simultaneously* fulfill the criteria for both a manic episode and for a major depressive episode (except for the duration criterion of two weeks) nearly every day during a period of at least one week. In reviewing these criteria, the Mood Disorders Workgroup of the DSM-5 Task Force concluded that while this definition of Mixed Episode had a satisfying symmetry, it was rather like a unicorn: beautiful to imagine, but rarely, if ever, seen in reality. Yet, the mixed episode diagnosis was actually recorded with substantial frequency and patients were very often referred to as being "mixed" in discussions among clinicians. Furthermore, we identified a number of negative consequences of the DSM-IV definition. These include (1) general confusion and lack of clarity about a patients actual clinical state; (2) underestimation of suicide risk, inasmuch as even softer definitions of mixed states are associated with increased risk of suicide; (3) potentially inappropriate treatment selection, given the poor response to lithium among patients in a mixed state; and, finally, (4) a failure to identify those with a lifetime diagnosis of unipolar disorder who are at increased risk of progression to bipolar disorder. We, therefore, set as our goal a redefinition of Mixed Episode that would better conform to clinical reality and that would allow for the recognition of mixed states even among individuals with a lifetime diagnosis of unipolar disorder.

The workgroup is currently proposing mixed states be identified by *specifiers* that can be applied to episodes of either depression or mania/hypomania and that can be applied to individuals with a lifetime diagnosis of either unipolar or bipolar disorder.

The proposal for the *with depressive features* specifier requires that full criteria are met for either a manic episode or a hypomanic episode and that at least three of the following nine depressive symptoms are present nearly every day during the episode: subjective depression, worry, self-reproach, or guilt, negative self-evaluation, hopelessness, suicidal ideation or behavior, anhedonia, fatigue, and psychomotor retardation.

The proposal for the *with hypomanic features* specifier requires that full criteria are met for a MDE and that at least three of the following seven manic/hypomanic symptoms are present nearly every day during the episode: elevated mood, decreased *need* for sleep (i.e., sleeps little, but reports feeling rested), increased goal-directed activity, increased energy/visible hyperactivity (as distinct from agitation), grandiosity, accelerated speech, and racing thoughts.

The literature on mixed states suggested that a number of the generally accepted validators of a diagnostic entity apply to the softer definitions proposed by the workgroup. These validators include familial aggregation of both predominantly manic and predominantly depressive mixed states, prior psychiatric history variables, concurrent psychiatric symptoms not included in the definition, diagnostic stability over time, increased likelihood of progression from a unipolar to a bipolar diagnosis among those with mixed depressive episodes and treatment outcome.

Features of psychiatric history that are consistently observed among those experiencing mixed states, such as those defined by the proposed mixed specifiers, include early onset of illness, a history of multiple previous episodes, suicidal behavior, comorbid diagnoses of anxiety and alcohol or substance abuse, and brain trauma. Interestingly, all of these are also features that tend to distinguish individuals with a lifetime history of bipolar disorder from those with a lifetime history of unipolar depression.

Two major concerns of the workgroup in developing the mixed specifier criteria were to select the most appropriate symptoms and to decide on the number of symptoms that should be required. In selecting the specific symptoms to be included, the workgroup focused on selecting those symptoms that were clearly distinct from those of the predominant episode polarity. This was done entirely based on face validity, as no literature was available to guide us. In deciding on the number of symptoms that should be required, we looked to previous studies of validators and to information about the likelihood of change in diagnosis. In a longitudinal follow-up study lasting almost 30 years, Fiedorowicz *et al.*[45] found a highly significant difference in the risk of conversion from unipolar to bipolar disorder among depressed individuals who presented with two vesus three or more manic/hypomanic symptoms.

The concept of Bipolar Disorder, Not Otherwise Specified (NOS) as it appeared in DSM-IV, and likely to be referred to as Bipolar Disorder, Not Elsewhere Classified (NEC) in DSM-5, will be reserved for presentations of bipolar disorder that are subsyndromal by virtue of an insufficient number of symptoms to meet the criteria for episodes of depression and/or mania or hypomania, or an insufficient episode duration to meet those criteria and there is no requirement that these subsyndromal presentations be concurrent. By contrast, to meet the criteria for one of the proposed mixed specifiers, an individual would be required to present in a fully syndromal episode of depression or mania or hypomania that is simultaneous for at least one week with at least 3 of the designated symptoms of the opposite pole.

Recognizing the considerable prognostic importance of mixed states, the DSM-5 workgroup on mood disorders has proposed a new definition of such states that we believe is more consistent with clinical reality, is likely to be used more appropriately and correctly than the mixed episode diagnosis in DSM-IV, should lead to earlier recognition of those depressed individuals likely to develop bipolar disorder and, perhaps most importantly, may help in the recognition of those at risk for suicidal behavior.

Trisha Suppes (Stanford University) finished this session with a third report. A debate often faced by psychiatrists working in mood disorders is what combination of symptoms is required to make a diagnosis of bipolar disorder, particularly which symptoms and for what duration. DSM-IV-TR offers guidance by distinguishing the criteria for number of hypomanic symptoms and concurrent days of hypomania and mania in bipolar disorder; however, there has been increasing recognition of both the presence and importance of subsyndromal hypomanic symptoms for patients experiencing distress.

The Mood Disorders Workgroup for the DSM-5 is making the recommendation for the addition of "activity or energy" to criterion A for hypomanic and manic episodes. The proposal for the updated criterion is as follows: a distinct period of abnormally and persistently elevated, expansive, or irritable mood and *abnormally and persistently increased activity or energy*. Symptom lists (criterion B) would be essentially not changed. The addition of requiring increased activity/energy to criterion A will make explicit the requirement that this hallmark symptom of bipolar I disorder be present in order to make the diagnosis. Addition of this feature is supported by a number of studies suggesting that activity or energy is at least as important as mood; moreover, some studies argue that activity and energy are more important than mood.[46–55]

The Mood Disorders Workgroup for DSM-5 is also making the recommendation to maintain the criterion A for an episode of hypomania at four days. The proposal is as follows: a distinct period of abnormally and persistently elevated, expansive, or irritable mood and abnormally and persistently increased activity or energy, *lasting at least four days and present most of the day*, nearly every day, that is clearly different from the usual nondepressed mood.

The evidence is fairly compelling that maintaining at least a four day requirement retains the specificity of the diagnosis with reasonable sensitivity. The review of the evidence suggests that decreasing the duration requirement would significantly increase the prevalence of bipolar II disorder. This could, in turn, decrease specificity of the diagnosis and add to the periodic skepticism regarding the integrity and validity of bipolar II as a separate diagnostic category. This recommendation is based on a review of available empirical data potentially supporting a change in the criteria, discussed in detail below. Under consideration was whether there was justification to change the criterion for a hypomanic episode to two days, shortening the required period of symptoms from four days. Some researchers have suggested that the duration requirement be eliminated. For example, Angst and colleagues,[47] as well as others, have suggested that the duration requirement should be discarded and the focus moved to changes in activity plus clinically significant symptoms. In one single-site observational study, characteristics of patients reporting two versus four day hypomania were similar.[56]

The recent BRIDGE study, is one of the few studies in which duration of hypomania is a specific focus. In this large globally ascertained sample of more than 5,500 patients currently experiencing a MDE, an evaluation was made of various cut points for duration between 1, 2–3, and 4–6 days of hypomania relative to a number of external validators.[42] The external validators considered included: first degree relatives with a mood disorder; early adult onset; recurrent mood episodes; mood lability; seasonality; and history of suicide attempts. There was a clear linear relationship between each of the external validators evaluated and the duration of the hypomanic symptoms. For example, the percent of the population reporting family history of mood disorders increased as number of days of hypomania increased. The difference between requiring 2–3 versus 4–6 in these validators was notable, in many cases improving by 20% or more.

In further support of maintaining duration criteria at four days, Bauer and colleagues[57] found that, in a sample of 203 bipolar patients, if the hypomania duration criterion was reduced from four days to two days, the percent of hypomanic days would increase twofold from 4% to 8% for each patient; the number of patients with a hypomanic episode would double (in this sample from 44 to 96); and the number of hypomanic episodes for all patients would increase about threefold (from 129 to 404 in this sample).

In sum, these data, particularly the increase in strength and number of external validators when a hypomania episode was defined by a minimum of four days,[42] and the likely marked increase in prevalence for bipolar II disorder that would occur if the entry criteria for hypomania were changed,[57] were taken by the Mood Disorder Workgroup committee to support maintaining the hypomania episode requirement at four days.

Finally, the Mood Disorders Workgroup for the DSM-5 proposes to revise Bipolar Disorder NOS to allow subcategories to be coded and to be specific in definition. This is an important change from DSM-IV-RS where NOS was defined by "examples include." This proposed change will allow the capture of the well-recognized phenomena of patients experiencing hypomanic symptoms with notable change from usual behavior that are fewer than four days in duration or of a lower symptom count than is required to meet criteria for a full hypomanic episode. It is important to note that in contrast to the Mixed Features Specifier proposed,[58] where symptoms occur simultaneously, symptoms for Bipolar NOS, by definition, occur at separate times. Specific proposed definitions include:

- Codable subcategories within Bipolar NOS that would classify characteristic symptoms of hypo/mania or depression that are present *during separate time periods* and cause distress or dysfunction, but are not of sufficient duration and/or intensity to meet criteria for a specific bipolar diagnosis.
- Subsyndromal hypomania, short duration: patients who experience lifetime episodes of depression that meet full criteria for MDE AND experience hypomanic periods of sufficient number of criterion symptoms, but of insufficient duration (≥ 2 and < 4 consecutive days).

- Subsyndromal hypomania, insufficient symptoms: patients who experience lifetime episodes of depression that meet full criteria for MDE AND experience hypomanic periods of sufficient duration, but of insufficient number of criterion symptoms (≥ 2, 3 consecutive days if mood is only irritable).
- Other Bipolar NOS: this includes atypical presentations of bipolar symptoms not considered above that cause significant distress or psychosocial dysfunction.
- In summary, in this session, DSM-5 Mood Disorder Workgroup proposals were reviewed, including the addition of activity or energy to the mood item in criterion A; that the duration criterion for hypomanic episodes remain at four days; and that specific, codable Bipolar NOS categories for subthreshold mood symptoms be added. More information about proposed changes in DSM-5 can be found at www.dsm5.org.

Session on medical comorbidity

Overview

In the comorbidity session chaired by Fouzia Laghrissi-Thode (F. Hoffmann-La Roche Ltd.), an emphasis was placed on the considerable medical comorbidity present in bipolar disorder. The burden of cardiometabolic conditions in bipolar disorder is vast and has an impact on levels of morbidity and mortality, and a reduced life expectancy of 10 to 25 years. In this session, there was the following: a review of treatment disparities; recommendations for integrative care; the examination of risk factors and how to address them; and, finally, the current state of intervention in an integrated care model. Several presentations addressed these important issues.

Cardiometabolic conditions

Benjamin Goldstein (University of Toronto) discussed the burden of cardiometabolic conditions in bipolar disorder. Cardiometabolic conditions, such as obesity, diabetes, and cardiovascular disease, are a common source of morbidity and mortality in bipolar disorder. Excessive cardiovascular disease (CVD) is the leading cause of death in bipolar disorder, contributing to 10- to 25-year shorter life expectancy. Despite the fact that CVD is the leading cause of death in the general population, standardized mortality ratios are fully doubled in bipolar

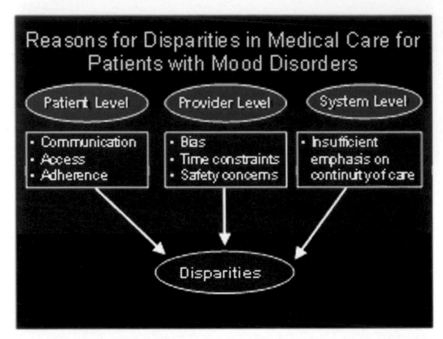

Figure 4. Reasons for disparities in medical care for patients with mood disorders, at the patient level, provider level, and system level. Adapted from Frayne *et al.*[121]

disorder, in large part because CVD in these patients occurs exceedingly early in life. Recent findings from the United States general population suggest that CVD patients with bipolar disorder are 14 years younger than CVD patients without mood disorders. Indeed, evidence of increased cardiovascular risk among youth with bipolar disorder is accumulating. The question of whether cardiometabolic conditions in bipolar disorder are due to the strain of having the illness (e.g., mood symptoms and their functional sequelae), suboptimal lifestyle (e.g., sleep disruption, smoking, nutrition, physical activity), or shared biological causes (e.g., inflammatory mediators, hypothalamic–pituitary–adrenal disturbance, CACNA1C genotype) remains insufficiently answered. Nonetheless, in addition to concerns about medical outcomes, cardiometabolic conditions are associated with a more pernicious course of bipolar disorder.[59]

Disparities exist in the medical treatment of patients with severe mental illness, and this has been observed in rigorous studies from a number of countries including Canada, Denmark, Sweden, and the United States. For example, despite being nearly three times as likely to die of CVD, patients with bipolar disorder or schizophrenia are no more likely

to access related hospital resources. Those who do access treatment are at greatly increased risk of cardiovascular death within five years, and yet are almost half as likely to receive invasive cardiovascular treatment as patients without these psychiatric illnesses. Other findings demonstrate extensive delays in treatment from the first point of contact through cardiac catheterization, and disparities in prescriptions for preventive cardiometabolic medications postdischarge have been documented. Potential reasons for these disparities have been highlighted by others (see Fig. 4). Cardiometabolic monitoring of second-generation antipsychotics is problematic. Despite publication of joint guidelines by the American Diabetes Association and American Psychiatric Association, adherence with suggested monitoring at baseline and after 12 weeks of treatment remains dismal, with the vast majority of patients not receiving monitoring. Among youth, adherence with monitoring guidelines occurs in only 5% to 15% of patients.

Strategies that may prevent or delay the accumulation of medical burden in bipolar disorder include regular medical monitoring, behavioral interventions focusing on obesity prevention, and use of medications that minimize the propensity for weight gain and metabolic disturbance. Treatment models for integrating greater emphasis on medical comorbidity in the treatment of patients with severe mental illness have recently been elaborated and published in landmark studies. Integrating nurse care managers into mental health clinics was shown to increase compliance with preventive services and evidence-based cardiometabolic services and to improve Framingham scores. Similarly, integrating supervised nurses into the provision of primary care for patients with depression and diabetes or cardiovascular disease led to more fine-tuned adjustments of insulin, antihypertensives, and antidepressants. Importantly, reductions in glycosylated hemoglobin and systolic blood pressure, and improvements in quality of life and treatment satisfaction, were also observed. Preliminary findings suggest that similar modifications can benefit the overall health of patients with bipolar disorder specifically. Although these manualized interventions are not widely available, they share several core elements that can be readily integrated into treatment. These include: psychoeducation about the link between psychiatric and cardiometabolic illness, liaison with other healthcare providers about specific patients and about monitoring and risk factor management in general, encouraging self management, and supporting patient self-efficacy and effective engagement of healthcare providers.

In summary, medical comorbidity is salient to the care of all patients with bipolar disorder, irrespective of age or type of treatment. Familiarity with the magnitude of this risk and with preventive and intervention strategies may improve outcomes for patients with bipolar disorder.

Sleep and circadian rhythms

In the second talk in the session, Allison Harvey (University of California at Berkeley) then presented an important overview of the role of sleep and circadian rhythms in bipolar disorder. Bipolar disorder is a severe and chronic psychiatric illness associated with significant interepisode impairment, high rates of medical morbidity, and premature mortality. Three lines of evidence highlight the strong coupling between sleep and circadian disturbances with mood episodes, inadequate recovery, and relapse risk in bipolar disorder: (1) sleep and circadian disturbance is a core symptom of bipolar disorder; (2)

experimental studies suggest that sleep deprivation can trigger manic relapse; and (3) there is evidence that sleep deprivation can adversely affect emotion regulation the following day.[60] Relatively recently, it has become clear that bipolar disorder is also associated with greater medical morbidity and premature mortality.[61] Multiple complex factors are likely to contribute to the association between bipolar disorder and health problems, medication side effects perhaps being the most prominent. The question we ask and seek to begin to answer here is: Are the sleep and circadian problems that are core features of bipolar disorder one important but understudied contributor to the known association between health problems and bipolar disorder? Below we seek to show the plausibility of this hypothesis by highlighting that problems relating to a selection of health conditions and health behaviors have been linked to both sleep problems and bipolar disorder. If sleep problems are contributors to the health problems experienced by bipolar patients, the public health implications are potentially startling because sleep and circadian problems are *modifiable*. Harvey summarized the following areas:

Several longitudinal studies have identified short sleep duration, long sleep duration, and insomnia as predictors of CVD morbidity and mortality.[62] Sleep disturbance has also been associated with the increased prevalence of traditional CVD risk factors, including obesity, hypertension, and diabetes mellitus[3].

There is a 1.5–2.5 fold increase in risk for CVD-related mortality for individuals with bipolar disorder compared to the general population, making CVD the leading cause of death in bipolar disorder.[63] Moreover, traditional CVD risk factors, such as hypertension, obesity, and diabetes mellitus, occur with greater frequency in bipolar patients than the general population.[64]

The results of a meta-analysis involving 30 studies (12 in children, 18 in adults) and 634,511 participants are compelling. There was a 60–80% increase in the odds of being a short sleeper among both adults and children who were obese. Moreover, adults sleeping five hours or fewer versus more than five hours had significantly greater odds of being obese and an increase of one hour per night of sleep was associated with a decrease of 0.35 body mass index (BMI).[65] Fagiolini *et al.*[66] reported that 68% of patients with bipolar disorder were overweight, with 32% meeting criteria for obesity (less than 20% of controls met criteria for obesity). These individuals suffered a range of poorer outcomes including shorter time to recurrence of an episode, particularly of depression, and had a greater number of previous episodes of depression and mania.[66] Similarly, McElroy *et al.*[67] reported that 58% of bipolar patients were overweight and 21% were obese. The associated adverse outcomes included arthritis, hypertension, and diabetes mellitus.

Sleep deprivation is associated with hormonal responses that increase appetite, caloric intake, and influence the selection of foods.[68] When sleep deprived, individuals focus food selection on sugar, fat, and carbohydrates. In a comparison of 2,032 patients with bipolar disorder and controls, Kilbourne *et al.*[69] reported that patients with bipolar disorder were more likely to report poorer eating behaviors relative to individuals without a serious mental illness, including having fewer than two daily meals (OR = 1.32) and difficulty obtaining or cooking food (OR = 1.48).

Sleep quality is improved by exercise[70] and is an effective treatment for chronic insomnia[71] and reduces presleep anxiety.[72] Moreover, healthy participants have less tolerance for exercise after sleep deprivation.[73] Kilbourne *et al.*[69] reported that patients with bipolar disorder were more likely to report poor exercise habits relative to individuals without a serious mental illness, including infrequent walking (OR = 1.33) and infrequent strength exercises (OR = 1.28). Another study reported predominately sedentary routine daily activities in a sample of bipolar patients.[74]

We have reviewed evidence that sleep disturbance is associated with an increased CVD risk, more obesity, poorer diet, and less exercise. We have also reviewed evidence that sleep disturbance is pervasive in bipolar disorder and that bipolar disorder is associated with higher CVD risk, more obesity, poorer diet, and less exercise. Taken together, it is tempting to speculate that sleep disturbance may be an important contributor to the association between health problems and bipolar disorder. However, there is surprisingly little research that includes a measure of sleep and a measure of a health outcome in individuals with bipolar disorder. Accordingly, we recently probed the National Comorbidity Survey-Replication (NCS-R) to examine the prevalence of three self-reported cardiovascular risk factors (obesity, hypertension, and diabetes) across bipolar respondents with chronic insomnia symptoms, acute insomnia symptoms, and good sleep ($n = 176$). Insomnia symptoms included difficulty falling asleep, difficulty maintaining sleep, and early morning awakening. Rates of obesity and hypertension were greater in bipolar patients with chronic (41.8%; 28.8%) and acute (43.7%; 24.1%) insomnia symptoms compared to bipolar patients with good sleep (19.7%; 5.9%). Longer insomnia symptom duration increased odds of hypertension (OR = 2.2, 95% CI = 1.2–3.9) and a higher number of insomnia symptoms was associated with elevated rates of obesity (OR = 1.5, 95% CI = 1.0–2.1), hypertension (OR = 1.9, 95% CI = 1.1–3.6) and diabetes (OR = 4.7, 95% CI = 2.0–11.1). Although causal inferences are not possible given the cross-sectional design, these results add to the evidence that sleep disturbance may contribute to the association between health problems and bipolar disorder.[75]

With funding from NIMH, we have developed an eight session intervention for sleep and circadian problems in bipolar disorder, combining principles from motivational interviewing,[76] cognitive behavior therapy for insomnia,[77] interpersonal and social rhythms therapy (IPSRT),[78] and chronotherapy.[79] The hypothesis tested is that improving sleep and circadian problems will improve sleep, reduce mood symptom and risk of relapse, improve quality of life as well as health among individuals with bipolar disorder.

Unlocking the contributors to the increased risk for premature health-related morbidity and morbidity associated with bipolar disorder will be complex as there will be multiple factors. Herein, we present initial evidence, albeit all based on cross-sectional evidence, that the sleep and circadian problems that are prominent features of bipolar disorder may contribute to several adverse health outcomes and may contribute to an unhealthy lifestyle. Sleep and circadian problems are likely to be *modifiable* with relatively simple behavioral changes. Hence, if sleep/circadian problems do turn out to be important contributors to the health problems experienced by bipolar patients, there will be an important opportunity for major health improvement.

Adverse health-related behaviors

In the final report of this session, Michael Ostacher (Stanford University) reported on several other adverse risk factors and behaviors that increase the morbidity and mortality in bipolar disorders. Adverse health-related behaviors increase the risk of morbidity in bipolar disorder. High rates of obesity, smoking, drug and alcohol use, sleep cycle abnormalities, and inadequate treatment adherence conspire to worsen the course of bipolar disorder for many people. Perhaps more importantly, these may be responsible for elevated mortality rates compared to the general population, mediated through diabetes, lipid disorders, hypertension, and heart disease. Clinicians should be aware of the presence of these risks, and address and promote behavioral change and optimization of care. Smoking, alcohol, diet, and exercise, and barriers to such changes should be addressed from the perspective of ongoing treatment. Strategies for assessing these risks, monitoring motivation to change behavior, and intervening to promote change must be integrated into patient care.

It has become evident that a singular focus on mood symptoms in the study and treatment of bipolar disorder neglects what may be the most important outcome we could measure: health. People with severe mental illness in the United States may die 25 years earlier than their counterparts in the general population. While suicide certainly accounts for a disproportionate number of deaths in the first decade or two after the development of bipolar disorder, standardized mortality ratios for people with bipolar disorder continue to be double that of the general population. Nearly all of this increased mortality is due to medical causes. While there may be something inherent in bipolar disorder that increases risk for premature mortality (abnormalities in inflammatory factors, for instance), it is likely that the greatest proportion of increased risk is due to modifiable health behaviors, primarily smoking and obesity. If patients are to be well treated in the clinic, these behaviors must be addressed.

Patients with bipolar disorder are at increased risk for multiple problems—elevated BMI, smoking, diabetes, lipid disorders—that individually increase risk for cardiovascular events, but having multiple risk factors, according to the Framingham Heart Study, more than additively increases such risk. In public sector patients in Massachusetts, marked increased risk for death from cardiovascular disorder was apparent in severely mentally ill patients, including those with bipolar disorder, as early as age 25.

The association between obesity and poor mood and functional outcomes, including suicide attempts, in bipolar disorder is well known, as is its prevalence. In the Pittsburgh sample, upwards of 35% of subjects with bipolar disorder had BMIs in the obese range (>30). With obesity directly related to rates of diabetes and lipid disturbances, its high prevalence in bipolar disorder is especially concerning. Because patients with mental illnesses, including bipolar disorder, often receive less comprehensive medical care than those without mental illness, the consequences of these risks may be greater.

Smoking, too, is highly prevalent in bipolar disorder, with odds ratios for smoking in bipolar I and II disorders 3.5 and 3.2, respectively. Quit rates are lower in bipolar disorder and, aside from its association with suicidal behavior, has to be considered in understanding why mortality rates in bipolar disorder are so high. Smoking cessation is rarely studied in bipolar disorder, and patients, clinicians, and families may be fatalistic about the utility of interventions to help patients quit. Many psychiatric clinicians are unaware of or do not help patients utilize established smoking cessation resources, such as pharmacological interventions, smoking cessation programs, and quit lines.

It is time for psychiatric care to go beyond a focus on symptoms and functioning to include a focus on overall health outcomes and decreased mortality for our patients. Systematic identification processes for risk factors for mortality need to be part of everyday care, just as systematic diagnosis of psychiatric illnesses are. Educational efforts for providers that emphasize the detrimental effects of smoking, obesity, diabetes, lipid abnormalities, and other health risk behaviors on both mental and physical health, and strategies (such as motivational interviewing) need to become part of practice for all psychiatric caregivers, and needs to include careful weighing of the potentially detrimental health effects of many of our medications. The integration of interventions for smoking cessation, dietary change, and weight management, overall physical health behavior change into ongoing care needs to be a primary goal of our field. Our technologies for improving mood and functioning are limited and improving only incrementally; now is the time to expect the view of psychiatric care to go beyond psychiatric symptoms and to expand to include overall health.

Session on reports from the International Society for Bipolar Disorder

Overview

The final session, coordinated by Michael Berk (University of Melbourne), discussed a series of reports from the International Society for Bipolar Disorders (ISBD). Bipolar disorder is a complex, common, and capricious disorder. Relative to its burden, which is between 1% and 5% of the global population, it is poorly researched and understood. Akin to what has been achieved in other major medical disorders, strategic international collaborations, networks, and partnerships are essential to leverage the resources and intellectual capital required to have an impact on these unmet needs. In this context, the series of special workgroups established by the ISBD attempts to address some of the core issues facing the field. The ISBD is the principal internationally recognized forum to foster ongoing international collaboration on education and research, with an objective to advance the treatment of all aspects of bipolar disorders, resulting in improvements in outcomes and quality of life for those with bipolar disorder and their significant others. In this context, the ISBD has chosen a number of areas for academic investment, including improving the quality and standardization of clinical trials, and fostering a better understanding and integration of the changes in cognition, neuroimaging, and biomarkers that are seen in the disorder.

The first issue pertains to the complexities and controversies regarding the design of clinical trials in bipolar disorders. The clinical trial evidence base is the cornerstone of clinical decision making, and the area is bedeviled by methodological inconsistency that bedevils quality care. Many trials fail because of poor design, and others contain critical biases. This workgroup, led by Tohen and colleagues will critically review the commonly utilized designs in clinical trials of bipolar disorder, as well as suggesting novel designs, including adaptive designs and mixed methods designs. A review of statistical techniques and cultural issues and challenges of implementing studies in emerging countries is part of the remit of this group. A major potential contribution to outcomes could derive from stan-

dardisation and optimisation of clinical trial design, and the adoption of innovative methodologies including statistical methods.

It is now appreciated that rather than simply being a disorder with highs and lows, with return to one's old self between episodes, bipolar disorder is associated with a cascade of progressive clinical features. Bipolar disorder follows a progressive trajectory; with an increased numbers of episodes and persistence of illness, there is an incrementally greater probability of recurrence, recurrence becomes triggered more easily, and there is a reduced likelihood of response to treatment. An active biological process of neuroprogression underpins this staged process, evidenced by novel evidence of both progressive neuroanatomical changes and cognitive decline. The biochemical foundations of this process appear to be mediated by changes in inflammatory cytokines, corticosteroids, neurotrophins and oxidative stress. The consequences of this noxious cascade include lipid peroxidation, protein carbonylation, DNA fragmentation and an increased vulnerability to apoptosis. This neural toxicity has overt functional consequences.

The Neurocognition Task Force, led by Frangou and Yatham, aims to examine cognitive deficits in bipolar disorder as part of the process of neuroprogression in the disorder and to develop potential therapeutic strategies. Key goals are to standardize cognitive tests that are commonly used, ensure that they are validated in patients with bipolar disorder, and confirm that they target those domains that are most relevant to the disorder. In order to establish a common cognitive battery for bipolar disorder, the ISBD put together a working group tasked with reviewing the cognition literature to propose a preliminary neurocognitive battery for use in bipolar disorder. The ISBD-Battery for Neurocognitive Assessment (ISDB-BANC) was selected as an assemblage of recognized individual tests considered appropriate for bipolar disorder. It initiates the procedure for standardizing cognitive testing in bipolar disorder, which is a critical step to the use of comparable metrics across large-scale clinical studies. This provides the foundation for the next generation of cognitive remediation therapies. It is planned that the findings from the various task force will be integrated with the expectation that new strategies for assessment and treatment can be developed in the near future.

The ISBD Neuroimaging Task Force similarly had two primary goals, to assimilate neuroimaging research from leading bipolar disorder groups and to develop a conceptual agreement regarding the functional neuroanatomy of bipolar disorder. The task force aimed to examine the neural systems that modulate mood, as they are the probable neuroanatomical foundation of bipolar disorder. Affected brain areas include the amygdala, pituitary, hippocampus, and disruption of regional white matter connectivity. Neuroimaging data indicate changes in key components of both structural and functional networks in bipolar disorder, both structural and functional. An important clue to the pathophysiology of the disorder rests in follow-up studies of the onset phase of the disorder, adolescence. There are normal developmental changes in neuroanatomy in adolescence, which overlap with the pathological changes that occur in bipolar disorder as part of the process of neuroprogression. These data dovetail with the work of the neurocognition taks force, providing a neuroanatomic substrate for the changes in functional outcome and cognition.

These structural and functional changes are the outcome of a primary biochemical perturbation. The root of these changes remain opaque, however biomarkers are increasingly illuminating the possibilities. To this end, the Biomarkers working group of the ISBD was set up, and the report by Young and Kapczinski highlights the most promising avenues. Oxidative stress and systemic inflammation appear to be key pathways underpinning neuroprogression, and they operate in concert with reductions in neurotrophins such as BDNF. The task group aims to develop large scale collaborative research linkages to examine the role of these biomarkers as moderators of outcome and of risk. Collaborative linkages such as these have a habit of growing organically, providing a critical mass of expertise to assist in developing solutions to complex issues of substantial public health significance.

Clinical trials task force report

In the first presentation in this session, Mauricio Tohen (University of Texas) summarized the report from the Clinical Trials Task Force. This group has focused on the review of commonly utilized designs in the treatment of bipolar disorder. The most common clinical trial design for acute mania and bipolar depression is a short, two arm study, which in general, can yield differences between efficacious interventions and placebo. As investigators, clinicians and regulatory authorities may recognize that symptomatic treatment of manic and depressive episodes is a relatively small component of management of bipolar disorder and, therefore, other designs need to be considered.

In general, monotherapy, single point randomization, and blinding achieve major aims. One important advantage of these designs is that they are relatively inexpensive and not subject to many ambiguities of interpretation. Additionally, it is possible to include an established efficacious treatment to provide "assay sensitivity" and allow for secondary analyses of comparative efficacy and tolerability between a new and an established drug/regimen. Designs that employ current DSM criteria and a total score on a manic rating scale have inherent weaknesses and are thus limited. DSM criteria treat all symptoms as equal, when evidence does support that conclusion. Similarly, total scores are subject to rater inflation to qualify individuals, thereby reducing study power to detect differences; as well as overweighting of nonspecific symptoms that may advantage one drug over another. For example, sedation alone can reduce many manic symptom scores, but the aim of treatment is rarely principally sedation.

A major unmet need is that little recognition has been given to the reality that most bipolar I patients have few full manic or depressive episodes while in treatment. Rather, less severe, and generally shorter periods of manic or depressive recrudescence occur more often. Yet, almost no systematic experimental studies have been conducted on this group.

In terms of maintenance or prophylactic studies, adaptive designs could include alternative randomized actions when manic symptoms worsen. Other investigative actions would proceed without change. The analyses of results would include additional hypotheses around the exacerbations in manic or depressive symptomatology

Special populations for bipolar studies fall into two categories. First are characteristics that are unlikely to have major impact on interventions and their effectiveness. These include ethnic and racial identity, sex, and socioeconomic status. However, this does not dismiss as without benefit some studies; for example, several studies have provided some evidence for greater prevalence/severity of psychotic symptoms among African American patients.

Quality of life is of limited relevance to bipolar studies in the short run but of major importance in maintenance stuies. Studies have consistently reported that functional and quality of life benefits lag behind recovery from syndromal clinical states in bipolar disorder. However, adequate and psychometrically sound measures of QOL exist, and have yielded differences among interventions. A general weakness in these and other ancillary outcome measures is that lack of adequate rater training, and often use of raters who are uninvolved in the clinical care of subjects can result in loss of sensitivity of such measures. Many of the comments about QOL apply similarly to measures of functioning.

Only recently have studies persuasively shown the plasticity of clinically significant cognitive function in bipolar disorder cognitive assessment tests. Generally, these assessments required time and effort that was off-putting to patients and expensive as a component cost of trials. Cognitive tests exhibiting sensitivity to changes with acute treatments are now available; therefore, the practical barrier to their more focused use is no longer paramount (see below).

Perhaps the major difference in emerging countries is that costs of certain regimens and drugs are prohibitive for most patients. Therefore, studies that compare drugs that are available generically—and their cost per year ranges—warrant more attention. Emerging economies are also less likely to have funding from federal revenues, therefore most potential studies in such countries will be ones funded by the pharmaceutical industry.

Neurocognition task force report

In the second presentation, Sophie Frangou (Kings College London) summarized the report from the Neurocognition Task Force. Cognitive impairment is part of the extended phenotype of bipolar disorder and is associated with genetic risk,[80] clinical severity,[81–86] and psychosocial outcome.[87] Examination of cognitive deficits in bipolar disorder is therefore integral to our efforts to define the pathophysiology of the disorder and to develop appropriate therapeutic strategies. Ideally, this should involve cognitive tests, which are standardized and validated in patients with bipolar disorder, targeting those domains that are most relevant to the disorder. A successful example of this approach is the Consensus Cognitive Battery (MCCB) developed by the Mea-

surement and Treatment Research to Improve Cognition in Schizophrenia (MATRICS) initiative.[88,89]

As a first step in establishing a common cognitive battery for bipolar disorder, the ISBD established a committee of experts who: (1) reviewed the literature to identify cognitive measures with high sensitivity for bipolar disorder and to identify gaps in current evidence; (2) evaluated the overlap with the MCCB and the usefulness of specific MCCB subtests for bipolar disorder; and (3) proposed a preliminary battery for use in bipolar disorder that also highlights areas where further research is required.

The literature review considered findings from meta-analyses of case-control studies of cognition in bipolar disorder[81–86] and from individual studies for specific cognitive domains with high relevance for bipolar disorder but with a limited evidence base. The evidence available is reasonably robust for some but not all cognitive domains of interest. Estimates of effect size are presented for those tests with sufficient information to allow meta-analytic treatment. Moderate to large impairments in bipolar disorder are present in tests of attention, processing speed, memory and learning, and in response inhibition and set-shifting. Remarkably, given the nature of the disorder, very few studies examined patients' performance in tests of emotional processing, social cognition, or decision making. This is, therefore, an area where further research is urgently required.

There is substantial empirical evidence suggesting that the MCCB subtests focused on the domains of attention and processing speed are suitable for bipolar disorder as these tests (or nearly identical versions) have already been demonstrated to show high sensitivity to the disorder. The MCCB subtests for the domains of working memory and visual learning/memory have not been as extensively studied in bipolar disorder, but available evidence suggests that they achieve a reasonable separation between bipolar disorder patients and controls, with effect sizes of 0.8 and 0.6–1, respectively.[90,91] Thus, it would appear reasonable at the present time to include these tests in a cognitive battery for bipolar disorder pending further confirmation by future studies. In the MCCB, verbal memory/learning is assessed with the Hopkins Verbal Learning Test (HVLT). This has not been a popular instrument in bipolar disorder because of concerns about its sensitivity. Indeed, case-control effect sizes for the HVLT in bipolar disorder range from 0.4–0.6;[90,91] these values are lower than

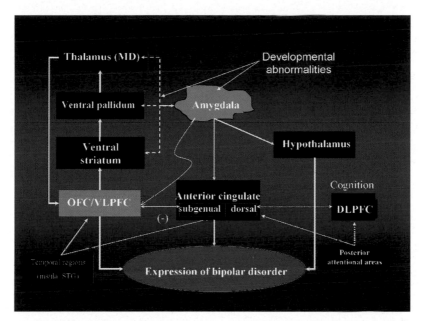

Figure 5. Schematic of abnormalities identified from neuroimaging within the ventromedial (OFC) and ventrolateral (VLPFC) emotional networks in bipolar disorder. Although the OFC and VLPFC are actually distinct circuits, they are combined here for simplicity. These networks modulate amygdala, which, in bipolar disorder, has functional and structural abnormalities, indicated by the uneven amygdala figure. Connections between amygdala and prefrontal cortex are abnormal as well, illustrated by the curving line. Prefrontal areas (OFC/VLPFC) are colored to indicate abnormalities in structure and function in these regions. Reciprocal connection between the emotional networks and cognitive (dorsal) brain systems, through the anterior cingulate, are indicated by the (−) on the arrow. Abbreviations: OFC, orbitofrontal cortex; VLPFC, ventrolateral prefrontal cortex; STG, superior temporal gyrus; DLPFC, dorsolateral prefrontal cortex.

the range typically reported in the meta-analytic studies summarizing data from more complex verbal list-learning measures, particularly the California Verbal Learning Test (CVLT). The CVLT may therefore be more appropriate for bipolar disorder patients, while the HVLT could be considered an acceptable substitute particularly in study designs that involve comparison with schizophrenia or repeated testing. The MCCB also targets reasoning and social cognition. Although both domains are relevant to bipolar disorder, the specific subtests have not been sufficiently studied or specifically implicated in bipolar disorder and are likely to lack assay sensitivity.[90] The greatest knowledge gap was identified in the assessment of the interaction between affective and nonaffective processes, where tasks of decision-making could prove particularly useful in bipolar disorder. Finally, after considering the validity and sensitivity of all the available tests discussed above, the committee also evaluated logistical aspects relating to brevity, ease of administration, reliability, repeatability, versatility and likely international utility.

The ISBD-Battery for Neurocognitive Assessment (ISDB-BANC) represents a collection of established individual tests deemed appropriate for use in bipolar disorder. It comprises eight core and two flexible subtests (choice between HVLT or CVLT) and, additionally, inclusion of the Wisconsin Card Sorting Test. The battery would be appropriate for applications where a broad screening of cognitive abilities is required, or where repeated assessments are necessary (e.g., treatment effects or clinical trials). Prospective studies in clinical populations will be required to support the psychometric properties of the ISBD-BANC and refine its constituent components. We consider the ISBD-BANC as the initial step in the process of standardizing cognitive testing in bipolar disorder.

Neuroimaging task force report

Steven Strakowski (University of Cincinnati) summarized the work of the Neuroimaging Task Force. The ISBD Neuroimaging Task Force met in Miami in December 2010 with two primary goals: (1) to review research from leading bipolar disorder

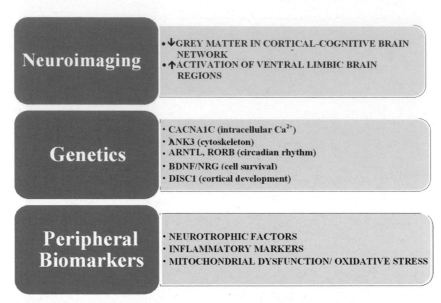

Figure 6. Potential candidates for biomarkers in bipolar disorder. Genetics studies have identified several potential candidate genes associated with increased risk for developing bipolar disorder, involving circadian rhythm, neuronal development, and calcium metabolism. Neuroimaging-based studies have consistently demonstrated loss of gray matter, and enlargement of striatum and amydgdala. In terms of peripheral biomarkers, repeated studies have found decreased levels of neurotrophic factors and increased proinflammatory cytokines and oxidative stress markers.

neuroimaging groups represented at the meeting; and (2) to develop consensus around a functional neuroanatomy for bipolar I disorder. The task force approached this subject working from the clinical assumption that bipolar disorder is a primary mood disorder; that is, that the systems most likely to underlie the condition involve those that modulate mood.

Although the specific control of emotional function in humans is not completely defined, two ventral prefrontal networks appear to modulate emotional behavior.[92–95] Both of these networks are similarly organized in that specific ventral prefrontal regions map to specific striatal then pallidal then thalamic brain areas to form iterative feedback loops that process information and modulate amygdala and other limbic brain areas. One network originates in the ventrolateral prefrontal cortex and is thought to modulate external emotional cues; the other originates in the ventromedial (orbitofrontal) cortex and is thought to modulate internal emotional stimuli.[92–95] These networks serve as a likely substrates for the functional neuroanatomy of bipolar disorder.

Neuroimaging studies suggest that abnormalities in key components of these networks occur in bipolar disorder. For example, Altshuler *et al.*[96] observed excessive amygdala activation in bipolar individuals during mania compared with healthy subjects while performing a facial affect matching task. Similar findings have been reported by others.[97–101] Abnormal amygdala activation has also been observed in bipolar disorder during other mood states.[102,103] Structural amygdala abnormalities are commonly reported in adults and adolescent bipolar subjects as well.[104,105]

A second common neuroimaging finding in bipolar disorder is abnormal ventral prefrontal activation. Several studies observed decreased activation in prefrontal cortex during mania,[90,106–108] which often occurs concurrently with amygdala overactivation.[109] Moreover, disrupted functional connectivity between amygdala and ventral prefrontal cortex has been observed during mania.[109] These findings suggest that loss of prefrontal modulation of amygdala activation may underlie the development of mood symptoms in bipolar disorder. Moreover, Strakowski *et al.* observed increased amygdala activation during euthymia that was associated with increased ventral prefrontal activation; the latter was interpreted to represent a compensatory mechanism in the "well" state to manage limbic brain overactivity. Several investigators have observed disruptions in the white matter connectivity between ventral prefrontal cortex and amygdala suggesting a structural basis for these functional observations.[110]

Bipolar disorder typically begins in adolescence. Recently, in a study of new-onset adolescent bipolar patients, during the year after the first manic episode, Bitter *et al.*[111] found that bipolar subjects did not exhibit amygdala growth that was seen in healthy adolescents (and adolescents with ADHD). However, at baseline, amygdala volumes among groups were the same, suggesting that amygdala developmental abnormalities occur as a result of illness progression, rather than as a cause of onset. In contrast, white matter abnormalities appear to precede the onset of bipolar illness as observed in studies of children at-risk for bipolar illness by virtue of having bipolar parents.[112]

Together, these data suggest a model of bipolar disorder (see Fig. 5) in which abnormalities within ventromedial and ventrolateral prefrontal networks lead to the expression of bipolar disorder. Disruptions in regional white matter connectivity may occur prior to illness onset, representing a potential vulnerability for developing bipolar illness. This model is illustrated in Figure 5 and provides a template to guide future investigations into the neurobiology of this important and common illness.

Biomarker committee report

Trevor Young (University of Toronto) summarized the report of the Biomarker group. The ISBD Biomarker Committee began meeting at the ICBD meeting in Pittsburgh in 2008, followed by gatherings at the ISBD meeting in San Paulo and several teleconferences. The Committee sought to have a comprehensive discussion of the area of biomarkers, which might help in the diagnosis and treatment of bipolar disorder. The group also took on a task to draft a positional paper which is under review at the journal *Bipolar Disorders*. The Committee met most recently at the ICBD meeting in Pittsburgh in June 2011.

It is clear that the etiology of bipolar disorder remains uncertain, however, there have been many recent advances in the area of genetics, pathophysiologic mechanisms, and brain imaging. The Committee reviewed the data and suggested that there were at least several biomarkers that could be identified from these three areas. The main potential candidates for biomarkers are shown in Figure 6. Neuroimaging studies have consistently shown loss of gray matter in cortical cognitive brain network[113] as well as alterations in the activation of relevant subfrontal, anterior temporal, and ventral prefrontal regions in response to an emotional stimulus in bipolar disorder.[114] Genetics has indicated several interesting potential candidate gene involvement ranging from circadian rhythms (ARNTL, RORB), to calcium metabolism (CACNA1C), cell survival, and cortical development (BDNF, DISC1).[97,115,116] With respect to peripheral biomarkers, three areas of particular interest include inflammation, mitochondrial dysfunction, and oxidative stress, and then finally changes in neurotrophic factors.[117,118]

Conclusions

Research in bipolar disorder has generated remarkable insights into the mechanisms associated with the disease risk, disease expression and treatment response. The insights highlight the complexity of bipolar disorder but also suggest that we may be close to successfully applying basic science findings in translational approaches to help us in our management of bipolar disorder. It is too early to know whether some of the most interesting targets identified will survive rigorous ongoing testing with adequate size samples and replication. To date, some of the biomarkers have already facilitated the development of new treatments, such as the work on using antioxidants in the treatment of symptoms of bipolar disorder. Techniques that strive to enhance brain health and to reduce the chance of cell loss have been a mainstay of bipolar disorder and are now supported by empirical work highlighting potential candidates that we may be able to measure in an ongoing fashion in patients.

Acknowledgments

The Ninth International Conference on Bipolar Disorders was supported in part by educational grants from AstraZeneca LP, Merck Sharp & Dohme Corporation, Bristol-Myers Squibb, and Cyberonics, Inc. Funding for this conference was also made possible by contributions from the Community Care Behavioral Health Organization (an affiliate of the UPMC Health Plan), the Fine Foundation, and the Staunton Farm Foundation.

Conflicts of interest

The authors declare no conflicts of interest.

References

1. Kato, T., C. Kakiuchi & K. Iwamoto. 2007. Comprehensive gene expression analysis in bipolar disorder. *Can J Psychiatry* **52**(12): 763–771.
2. Konradi, C., M. Eaton, M.L. MacDonald, *et al.* 2004. Molecular evidence for mitochondrial dysfunction in bipolar disorder. *Arch Gen Psychiatry* **61**(3): 300–308.
3. Sun, X., J.F. Wang, M. Tseng & L.T. Young. 2006. Down-regulation in components of the mitochondrial electron transport chain in the postmortem frontal cortex of subjects with bipolar disorder. *J Psychiatry Neurosci* **31**(3): 189–196.
4. Beneyto, M., E. Sibille & D.A. Lewis. 2008. Human post-mortem brain research in mental illness syndromes. In: D.S. Charney & E.J. nestler (eds). *Neurobiology of Mental Illness*. Oxford University Press.
5. Gaiteri, C., J.P. Guilloux, D.A. Lewis & E. Sibille. 2010. Altered gene synchrony suggests a combined hormone-mediated dysregulated state in major depression. *PLoS One* **5**(4): e9970.
6. Lee, H.K., A.K. Hsu, J. Sajdak, *et al.* 2003. Coexpression analysis of human genes across many microarray data sets. *Genome Research* **14**(6): 1085–1094.
7. Dobrin, R., J. Zhu, C. Molony, *et al.* 2009. Multi-tissue coexpression networks reveal unexpected subnetworks associated with disease. *Genome Biol.* **10**(5): R55.
8. Elo, L.L., H. Jarvenpaa, M. Oresic, *et al.* 2007. Systematic construction of gene coexpression networks with applications to human T helper cell differentiation process. *Bioinformatics* **23**(16): 2096–2103.
9. Bullmore, E. & O. Sporns. 2009. Complex brain networks: graph theoretical analysis of structural and functional systems. *Nat Rev Neurosci.* **10**(3): 186–198.
10. Gaiteri, C. & E. Sibille. Differentially-expressed genes in major depression reside on the periphery of resilient gene coexpression networks. *Front Neurol Neurosci*, in press.
11. Goh, K.I., M.E. Cusick, D. Valle, *et al.* 2007. The human disease network. *Proc Natl Acad Sci USA.* **104**(21): 8685–8690.
12. Huang, T.L. & F.C. Lin. 2007. High-sensitivity C-reactive protein levels in patients with major depressive disorder and bipolar mania. *Prog Neuropsychopharmacol Biol Psychiatry* **31**(2): 370–372.
13. Knijff, E.M., C. Ruwhof, H.J. de Wit, *et al.* 2006. Monocyte-derived dendritic cells in bipolar disorder. *Biol Psychiatry* **59**: 317–326.
14. Liu, H.C., Y.Y. Yang, Y.M. Chou, *et al.* 2004. Immunologic variables in acute mania of bipolar disorder. *Neuroimmunol* **150**: 116–122.
15. O'Brien S.M., P. Scully, L.V. Scott & T.G. Dinan. 2006. Cytokine profiles in bipolar affective disorder: Focus on acutely ill patients. *J Affect Disord* **90**: 263–267.
16. Wadee, A.A., R.H. Kuschke, L.A. Wood, *et al.* 2002. Serological observations in patients suffering from acute manic episodes. *Hum Psychopharmacol* **17**: 175–179.
17. Dickerson, F., C. Stallings, A. Origoni, *et al.* 2007. Elevated serum levels of C-reactive protein are associated with mania symptoms in outpatients with bipolar disorder. *Prog Neuropsychopharmacol Biol Psychiatry* **31**(4): 952–955.
18. Young, R.C., J.T. Biggs, V.E. Ziegler & D.A. Meyer. 1978. A rating scale for mania: reliability, validity and sensitivity. *Br J Psychiatry* **133**: 429–435.
19. Kay, S.R., A. Fiszbein & L.A. Opler. 1987. The positive and negative syndrome scale (PANSS) for schizophrenia. *Schizophr Bull.* **13**: 261–276.
20. Ortiz-Domínguez, A., M.E. Hernández, C. Berlanga, *et al.* 2007. Immune variations in bipolar disorder: phasic differences. *Bipolar Disord* **9**(6): 596–602.
21. Abeer, A. El-Sayed & H.A. Ramy. 2006. Immunological changes in patients with mania: changes in cell mediated immunity in a sample from Egyptian patients. *Egypt J. Immunol.* **13**(1): 79–85.
22. Cassidy, F. & B.J. Carroll. 2002. Hypocholesterolemia during mixed manic episodes. *Eur Arch Psychiatry Clin Neurosci* **252**(3): 110–114.
23. McMahon F.J., O.C. Stine, D.A. Meyers, *et al.* 1995. Patterns of maternal transmission in bipolar affective disorder. *Am J Hum Genet* **56**(6): 1277–1286.
24. Kato, T., O.C. Stine, F.J. McMahon & R.R. Crowe. 1987. Increased levels of a mitochondrial DNA deletion in the brain of patients with bipolar disorder. *Biol Psychiatry* **43**(10): 871–875.
25. C. Konradi. 2005. Gene expression microarray studies in polygenic psychiatric disorders: applications and data analysis. *Brain Res Brain Res Rev.* **50**(1): 142–155.
26. Andreazza, A.C., L. Shao, J.F. Wang & L.T. Young. 2010. Mitochondrial complex 1 activity and oxidative damage to mitochondrial proteins in the prefrontal cortex of patients with bipolar disorder. *Arch Gen Psychiatry* **67**(4): 360–368.
27. Gawryluk, J.W., J.F. Wang, A.C. Andreazza, *et al.* 2011. Decreased levels of glutathione, the major brain antioxidant, in post-mortem prefrontal cortex from patients with psychiatric disorders. *Int J Neuropsychopharmacol* **14**(1): 123–130.
28. Caltaldo, A.M., D.L. McPhie, N.T. Lange, *et al.* 2010. Abnormalities in mitochondrial structure in cells from patients with bipolar disorder. *Am J Pathol.* **177**(2): 575–585.
29. Andreazza, A.C., M. Kauer-Sant'anna, B.N. Frey, *et al.* 2008. Oxidative stress markers in bipolar disorder: a meta-analysis. *J Affect Disord* **111**(2–3): 135–144.
30. Wang, J.F. 2007. Defects of mitochondrial electron transport chain in bipolar disorder: Implications for mood-stabilizing treatment. *Can J Psychiatry* **52**(12): 753–762.
31. Berk, M., D.L. Copology, O. Dean, *et al.* 2008. N-acetyl cysteine for depressive symptoms in bipolar disorder – a double-blind randomized placebo-controlled trial. *Biol Psychiatry* **64**(6): 468–475.
32. Kraepelin, E. 1989. *Manic-Depressive Insanity and Paranoia*. Birmingham, Classics of Medicine Library.
33. McElroy S. 1992. Clinical and research implications of the diagnosis of dysphoric or mixed mania or hypomania. *Am J Psych.* **149**: 1633–1644.

34. Suppes P. Proposals to the DSM-5 for bipolar disorder: Hypomania duration and the emphasis on increased activity/energy. *Ann NY Acad Sci.*

35. Angst, J. 2007. The bipolar spectrum. *Br J Psychiatry* **190:** 189–191.

36. Cassano, G.B., P. Rucci, E. Frank, *et al.* 2004. The mood spectrum in unipolar and bipolar disorder: arguments for a unitary approach. *Am J Psychiatry* **161:** 1264–1269.

37. Hirschfeld, R.M.A., A.H. Young, K.D. Wagner, *et al.* 2004. *Screening for bipolar disorder in UK adults using the mood disorders questionnaire.* (Poster), XXIV CINP Congress. Paris.

38. Hantouche, E., J.M. Azorin, S. Lancrenon, *et al.* 2009. Prévalence de l'hypomanie dans les dépressions majeures récurrentes ou résistantes: enquêtes Bipolact. *Ann Méd Psychol.* **167:** 30–37.

39. Rybakowski, J.K., J. Angst, D. Dudek, *et al.* 2010. Polish version of the Hypomania Checklist (HCL-32) scale: the results in treatment-resistant depression. *Eur Arch Psychiat Neurol Sci.* **260:** 139–144.

40. Angst, J., L. Cui, J. Swendsen, *et al.* 2010. Major depressive disorder with subthreshold bipolarity in the National Comorbidity Survey Replication. *Am J Psychiatry* **167:** 1194–1201.

41. Zimmerman, P., T. Brückl, A. Nocon, *et al.* 2009. Heterogeneity of DSM-IV major depressive disorder as a consequence of subthreshold bipolarity. *Arch Gen Psychiatry* **66**(12): 1341–1352.

42. Angst, J., J.M. Azorin, C.L. Bowden, *et al.* 2011. Prevalence and characteristics of undiagnosed bipolar disorders in patients with a major depressive episode: the Bridge Study. *Arch Gen Psychiatry* **68:** 791–799.

43. Angst, J. 2009. Diagnostic concepts of bipolar disorders: a European perspective. *Clin Psychol Sci Prac.* **16:** 161–165.

44. Vieta, E. & M.L. Phillips. 2007. Deconstructing bipolar disorder: A critical review of its diagnostic validity and a proposal for DSM-V and ICD-11. *Schizophr Bull.* **33,** 886–892.

45. Fiedorowicz, J.G., J. Endicott, A.C. Leon, *et al.* 2011. Subthreshold hypomanic symptoms in progression from unipolar major depression to bipolar disorder. *Am J Psychiatry* **16:** 40–48.

46. Akiskal, H.S., E.G. Hantouche, M.L. Bourgeois, *et al.* 2001. Toward a refined phenomenology of mania: combining clinician-assessment and self-report in the French EPIMAN study. *J Affect Disord.* **67:** 89–96.

47. Angst, J., A. Gamma, F. Benazzi, *et al.* 2003. Toward a redefinition of subthreshold bipolarity: epidemiology and proposed criteria for bipolar-II, minor bipolar disorders and hypomania. *J Affect Disord.* **73:** 133–146.

48. Angst, J., A. Gamma, J-M. Azorin, *et al.* 2011. *Broad considerations of Bipolar I and Bipolar II Disorder.* International Conference on Bipolar Disorder (ICBD), Symposium II DSM-5. Pittsburgh, PA, June 9–11.

49. Bauer, M.S., P. Critis-Christoph, W.A. Ball, *et al.* 1991. Independent Assessment of Manic and Depressive Symptoms by Self-Rating: Scale characteristics and implications for the study of mania. *Arch Gen Psychiatry* **48:** 807–812.

50. Benazzi F. 2007. Testing new diagnostic criteria for hypomania. *Ann Clin Psychiatry* **19**(2): 1–6.

51. Benazzi, F. & H.S. Akiskal. 2003. The dual factor structure of self-rated MDQ hypomania: energized-activity versus irritable-thought racing. *J Affect Disord* **73:** 59–64.

52. Cassano, G.B., M. Mula, P. Rucci, *et al.* 2009. The Structure of lifetime manic-hypomanic spectrum. *J Affect Disord* **112:** 59–70.

53. Cassidy, F., E. Murry, K. Forest & B.J. Carroll. 1998. A factor analysis of the signs and symptoms of mania. *Arch Gen Psychiatry* **55:** 27–32.

54. Gardner, R. Jr & B. Wenegrat. 1991. What is the 'Core' symptom of mania? *Arch Gen Psychiatry* **50:** 71–72.

55. Goodwin, F.K. & K.R. Jamison. 2007. *Manic-Depressive Illness: Bipolar Disorder and Recurrent Depression.* Oxford University Press. Oxford, NY.

56. Benazzi, F. & H. Akiskal. 2006. The duration of hypomania in bipolar-II disorder in private practice: methodology and validation. *J Affect Disord* **96:** 189–196.

57. Bauer, M., P. Grof, N.L. Rasgon, *et al.* 2006. Self-reported data from patients with bipolar disorder: Impact on minimum episode length for hypomania. *J Affect Disord* **96**(1–2): 101–105.

58. Frank, E. 2011. Proposed revisions to the concept of mixed episodes in DSM-5: The path travelled. International Conference on Bipolar Disorder (ICBD), Symposium II DSM-5. Pittsburgh, PA, June 9–11.

59. Goldstein, B.I., A. Fagiolini, P. Houck & D.J. Kupfer. 2009. Cardiovascular disease and hypertension among adults with bipolar I disorder in the United States. *Bipolar Disord* **11**(6): 657–662.

60. Harvey, A.G., D.A. Schmidt, A. Scarna, *et al.* 2005. Sleep-related functioning in euthymic patients with bipolar disorder, patients with insomnia, and subjects without sleep problems. *Am J Psychiatry* **162:** 50–57.

61. Kupfer, D.J. 2005. The Increasing Medical Burden in Bipolar Disorder. *JAMA* **293:** 2528–2530.

62. Knutson, K.L. 2010. Sleep duration and cardiometabolic risk: A review of the epidemiologic evidence. *Best Prac Res Clin Endocrinol Metab.* **24:** 731–743.

63. Osby, U., K. Brandt, N. Correia, *et al.* 2001. Excess mortality in bipolar and unipolar disorder in Sweden. *Arch Gen Psychiatry* **58:** 844–850.

64. Birkenaes, A.B., S. Opjordsmoen & C. Brunborg. 2007. The level of cardiovascular risk factors in bipolar disorder equals that of schizophrenia: a comparative study. *Journal of Clinical Psychiatry* **68:** 917–923.

65. Cappuccio, F.P., F.M. Taggart, N.B. Kandala, A. Currie, E. Peile, S. Stranges & M.A. Miller. 2008. Meta-analysis of short sleep duration and obesity in children and adults. *Sleep* **31:** 619–626.

66. Fagiolini, A., D.J. Kupfer, P.R. Houck, *et al.* 2003. Obesity as a correlate of outcome in patients with bipolar I disorder. *Am J Psychiatry* **160:** 112–117.

67. McElroy S.L., M.A. Frye, T. Suppes, *et al.* 2002. Correlates of overweight and obesity in 644 patients with bipolar disorder. *J Clin Psychiatry* **63:** 207–213.

68. Spiegel, K., E. Tasali, P. Penev & E.V. Cauter. 2004. Brief communication: Sleep curtailment in healthy young men

is associated with decreased leptin levels, elevated ghrelin levels, and increased hunger and appetite. *Ann Intern Med.* **141:** 846–850.

69. Kilbourne, A.M., D.L. Rofey, J.F. McCarthy, *et al.* 2007. Nutrition and exercise behavior among patients with bipolar disorder. *Bipolar Disord.* **9:** 443–452.

70. M. Dworak, A. Wiater, D. Alfer, *et al.* 2008. Increased slow wave sleep and reduced stage 2 sleep in children depending on exercise intensity. *Sleep Medicine* **9:** 266–272.

71. Reid, K.J., K.G. Baron, B. Lu, *et al.* 2010. Aerobic exercise improves self-reported sleep and quality of life in older adults with insomnia. *Sleep Med.* **11:** 934–940.

72. G.S. Passos, D. Poyares, M.G. Santana, *et al.* 2010. Effect of acute physical exercise on patients with chronic primary insomnia. *J Clin Sleep Med.* **6:** 270–275.

73. Martin, B.J. 1981. Effect of sleep deprivation on tolerance of prolonged exercise. *Eur J Appl Physiol* **47:** 345–354.

74. Chuang, H.T., C. Mansell & S.B. Patten. 2008. Lifestyle characteristics of psychiatric outpatients. *Can J Psychiatry* **53:** 260–266.

75. Soehner A. & A.G. Harvey. 2011. Insomnia symptoms and cardiovascular risk in bipolar disorder: A national comorbidity survey-replication analysis. *Manuscript submitted for publication.*

76. W.R. Miller & S. Rollnick. 2002. Motivational interviewing: Preparing people to change. *Guilford Press.*

77. Morin, C.M. & C.A. Espie. 2003. Insomnia: A clinical guide to assessment and treatment. Kluwer Academic/Plenum Publishers.

78. E. Frank, D.J. Kupfer, M.E. Thase, *et al.* 2005. Two-Year Outcomes for Interpersonal and Social Rhythm Therapy in Individuals With Bipolar I Disorder. *Arch Gen Psychiatry* **62:** 996–1004.

79. Wirz-Justice A., F. Benedetti & M. Terman. 2009. Chronotherapeutics for affective disorders: A clinician's manual for light & wake therapy. Karger.

80. D.C. Glahn, L. Almasy, M. Barguil, *et al.* 2010. Neurocognitive endophenotypes for bipolar disorder identified in multiplex multigenerational families. *Arch Gen Psychiatry* **67:** 168–177.

81. Arts, B., N. Jabben, L. Krabbendam & van Os J. 2008. Meta-analyses of cognitive functioning in euthymic bipolar patients and their first-degree relatives. *Psychol Med.* **38:** 771–785.

82. Bora, E., M. Yucel & C. Pantelis. 2009. Cognitive endophenotypes of bipolar disorder: A meta-analysis of neuropsychological deficits in euthymic patients and their first-degree relatives. *J Affect Disord* **113:** 11–20.

83. Quraishi, S. & S. Frangou. 2002. Neuropsychology of bipolar disorder: a review. *J Affect Disord* **72:** 209–226.

84. Robinson, L.J., J.M. Thompson, P. Gallagher, *et al.* 2006. A meta-analysis of cognitive deficits in euthymic patients with bipolar disorder. *J Affect Disord* **93:** 105.

85. Stefanopoulou, E., A. Manoharan, S. Landau, *et al.* 2009. Cognitive functioning in patients with affective disorders and schizophrenia: A meta-analysis. *Int Rev Psychiatry* **21:** 336–56.

86. Torres, I.J., V.G. Boudreau & L.N. Yatham. 2007. Neuropsychological functioning in euthymic bipolar disorder: a meta-analysis. *Acta Psychiatr Scand Suppl.* **434:** 17–26.

87. Malhi, G.S., B. Ivanovski, Hadzi-Pavlovic D., *et al.* 2007. Neuropsychological deficits and functional impairment in bipolar depression, hypomania and euthymia. *Bipolar Disord* **9:** 114–125.

88. Green, M.F. & K.H. Nuechterlein. 2004. The MATRICS initiative: developing a consensus cognitive battery for clinical trials. *Schizophr Res.* **72:** 1–3.

89. Kern, R.S., J.M. Gold, D. Dickinson, *et al.* 2011. The MCCB impairment profile for schizophrenia outpatients: results from the MATRICS psychometric and standardization study. *Schizophr Res.* **126:** 124–131.

90. Burdick, K.E., T.E. Goldberg, B.A. Cornblatt, *et al.* 2011. The MATRICS consensus cognitive battery in patients with bipolar I disorder. *Neuropsychopharmacology* **36:** 1587–1592.

91. Schretlen, D.J., N.G. Cascella, S.M. Meyer, *et al.* 2007. Neuropsychological functioning in bipolar disorder and schizophrenia. *Biol Psychiatry* **62:** 179–186.

92. Chen, Y.C., D. Thaler, P.D. Nixon, *et al.* 1995. The functions of the medial premotor cortex. II. The timing and selection of learned movements. *Exp Brain Res.* **102:** 461–473.

93. Lane, R.D., E.M. Reiman, B. Axelrod, *et al.* 1998. Neural correlates of levels of emotional awareness. Evidence of an interaction between emotion and attention in the anterior cingulate cortex. *J Cogn Neurosci.* **10:** 525–535.

94. Phan, K.L., T. Wager, S.F. Taylor & I. Liberzon. 2002. Functional neuroanatomy of emotion: a meta-analysis of emotion activation studies in PET and fMRI. *Neuroimage* **16:** 331–348.

95. Yamasaki, H., K.S. LaBar & G. McCarthy. 2002. Dissociable prefrontal brain systems for attention and emotion. *PNAS* **99:** 11447–11451.

96. Altshuler L., S. Bookheimer, M.A. Proenza, *et al.* 2005. Increased amygdala activation during mania: a functional magnetic resonance imaging study. *Am J Psychiatry* **162:** 1211–1213.

97. Almeida, J.R., A. Versace, S. Hassel, *et al.* 2010. Elevated amygdala activity to sad facial expressions: a state marker of bipolar but not unipolar depression. *Biol Psychiatry* **67**(5): 414–421.

98. Blumberg, H.P., N.H. Donegan, C.A. Sanislow, *et al.* 2005. Preliminary evidence for medication effects on functional abnormalities in the amygdala and anterior cingulate in bipolar disorder. *Psychopharmacology* **183**(3): 308–313.

99. Kalmar, J.H., F. Wang, L.G. Chepenik, *et al.* 2009. Relation between amygdala structure and function in adolescents with bipolar disorder. *J Am Acad Child Adolesc Psychiatry* **48:** 636–642.

100. Lawrence, N.S., A.M. Williams, S. Surguladze, *et al.* 2004. Subcortical and ventral prefrontal cortical neural responses to facial expressions distinguish patients with bipolar disorder and major depression. *Biol Psychiatry* **55**(6): 578–587.

101. Rich, B.A. & D.T. Vinton, R. Roberson-Nay, *et al.* 2006. Limbic hyperactivation during processing of neutral facial

expressions in children with bipolar disorder. *Proc Natl Acad Sci USA* **103**(23): 8900–8905.

102. Altshuler, L., S. Bookheimer, J. Townsend, *et al.* 2008. Regional brain changes in bipolar I depression: a functional magnetic resonance imaging study. *Bipolar Disord* **10:** 708–717.

103. Strakowski, S.M., C.M. Adler, S.K. Holland, *et al.* 2004. A preliminary FMRI study of sustained attention in euthymic, unmedicated bipolar disorder. *Neuropsychopharmacol* **29**(9): 1734–1740.

104. Pfeifer, J.C., J. Welge, S.M. Strakowski, *et al.* 2008. Meta-analysis of amygdala volumes in children and adolescents with bipolar disorder. *J Am Acad Child Adolesc Psychiatry* **47**(11): 1289–1298.

105. Strakowski, S.M., M.P. Delbello & C.M. Adler. 2005. The functional neuroanatomy of bipolar disorder: A review of neuroimaging findings. *Mol Psychiatry* **10**(1): 105–116.

106. Blumberg, H.P., E. Stern, S. Ricketts, *et al.* 1999. Rostral and orbital prefrontal cortex dysfunction in the manic state of bipolar disorder. Am J Psychiatry **156**(12): 1986–1988.

107. Blumberg, H.P., H.C. Leung, P. Skudlarski, *et al.* 2003. A functional magnetic resonance imaging study of bipolar disorder: state- and trait-related dysfunction in ventral prefrontal cortices. *Arch Gen Psychiatry* **60**: 601–609.

108. Strakowski, S.M., J.C. Eliassen, M. Lamy, M.A. Cerullo, J.B. Allendorfer, M. Madore, J.H. Lee, J.A. Welge, M.P. DelBello, D.E. Fleck & C.M. Adler. 2011. Functional magnetic resonance imaging brain activation in bipolar mania: evidence for disruption of the ventrolateral prefrontal-amygdala emotional pathway. *Biol Psychiatry* **69**(4): 381–388.

109. Foland, L.C., L.L. Altshuler, S.Y. Bookheimer, *et al.* 2008. Evidence for deficient modulation of amygdala response by prefrontal cortex in bipolar mania. *Psychiatry Res* **162**(1): 27–37.

110. Versace, A., J.R.C. Almeida, S. Hassel, *et al.* 2008. Elevated left and reduced right orbitomedial prefrontal fractional anisotropy in adults with bipolar disorder revealed by tract-based spatial statistics. *Arch Gen Psychiatry* **65**(9): 1041–1052.

111. Bitter, S.M., N.P. Mills, C.M. Adler, S.M. Strakowski & M.P. DelBello. Progression of Amygdala Volumetric Abnormalities in Adolescents After Their First Manic Episode. *J Am Acad Child Adolesc Psychiatry*, in press.

112. Versace, A., C.D. Ladouceur, S. Romero, *et al.* 2010. Altered development of white matter in youth at high familial risk for bipolar disorder: a diffusion tensor imaging study. *J Am Acad Child Adolesc Psychiatry* **49**(12): 1249–1259.

113. Adler, C.M., J. Adams, M.P. DelBello, *et al.* 2006. Evidence of white matter pathology in bipolar disorder adolescents experiencing their first episode of mania: a diffusion tensor imaging study. *Am J Psychiatry* **163**(2): 322–324.

114. Adler, C.M., S.K. Holland, V. Schmithorst, *et al.* 2004. Abnormal frontal white matter tracts in bipolar disorder: a diffusion tensor imaging study. *Bipolar Disord* **6**(3): 197–203.

115. Andreazza, A.C., C. Cassini, A.R. Rosa, *et al.* 2007. Serum S100B and antioxidant enzymes in bipolar patients. *J Psychiatr Res.* **41**(6): 523–529.

116. Andreazza, A.C., M. Kauer-Sant'anna, B.N. Frey, *et al.* 2008. Oxidative stress markers in bipolar disorder: a meta-analysis. *J Affect Disord* **111**(2–3): 135–144.

117. Athanasiu, L., M. Mattingsdal, A.K. Kahler, *et al.* 2010. Gene variants associated with schizophrenia in a Norwegian genome-wide study are replicated in a large European cohort. *J Psychiatr Res.* **44**(12): 748–753.

118. Baumer, F.M., M. Howe, K. Gallelli, *et al.* 2006. A pilot study of antidepressant-induced mania in pediatric bipolar disorder: Characteristics, risk factors, and the serotonin transporter gene. *Biol Psychiatry* **60**(9): 1005–1012.

119. Sibille, E., Y. Wang, J. Joeyen-Waldorf, *et al.* 2009. A molecular signature of depression in the amygdala. *Am. J. Psychiatry* **166**: 1011–1024.

120. Maletic, V., M. Robinson, T. Oakes, S. Iyengar, S.G. Ball & J. Russell. 2007. Neurobiology of depression: an integrated view of key findings. *Int. J. Clin. Pract.* **61**: 2030–2040.

121. Frayne, S.M., J.H. Halanych & D.R. Miller. 2005. Disparities in diabetes care: impact of mental illness. *Arch. Intern. Med.* **165**: 2631–2638.

Ann. N.Y. Acad. Sci. ISSN 0077-8923

ANNALS OF THE NEW YORK ACADEMY OF SCIENCES
Issue: Annals *Meeting Reports*

Sixth International Congress on Shwachman-Diamond syndrome: from patients to genes and back

Johnson M. Liu,[1,2] Jeffrey M. Lipton,[1,2] and Sridhar Mani[3]

[1]The Feinstein Institute for Medical Research, North Shore-Long Island Jewish Health System, New York. [2]Steven and Alexandra Cohen Children's Medical Center of New York, North Shore-Long Island Jewish Health System, New Hyde Park, New York. [3]Albert Einstein College of Medicine of Yeshiva University, Bronx, New York

Address for correspondence: Johnson M. Liu, M.D., The Feinstein Institute for Medical Research, 350 Community Drive, Manhasset, New York, 11030. JLiu3@NSHS.edu.

At the Sixth International Congress on Shwachman-Diamond syndrome, held at the New York Academy of Sciences on June 28–30, 2011, researchers from around the world met to discuss the latest clinical and basic science relating to this puzzling condition.

Keywords: Schwachman-Diamond syndrome; genetics; autosomal recessive

Overview

In 1964, two groups of physicians, led by Shwachman and Diamond[1] and by Bodian,[2] independently reported the discovery of a new syndrome in humans. Shwachman-Diamond syndrome (SDS) is a rare, autosomal recessive disorder that is usually identified in infancy and is characterized by exocrine pancreatic insufficiency and bone marrow failure often associated with neurodevelopmental and skeletal abnormalities (Fig. 1). Additional clinical features include short stature, growth failure, immune dysfunction, cardiac and renal defects, liver disease, diabetes mellitus, and a predisposition to malignant myeloid transformation during the second and third decade of life (Fig. 1). Almost 40 years after its initial description, SDS was linked to a genetic locus on chromosome 7 (ref 3), and subsequent studies identified mutations in a single gene, dubbed *SBDS*, in most, but not all, SDS patients.[4]

The Sixth International Congress on Shwachman-Diamond syndrome began with presentations about the major clinical features of SDS (Session I) and the recent development of working guidelines for its diagnosis and management. Because not all SDS patients have identifiable *SBDS* mutations, gene-based tests are sometimes insufficient, and the diagnosis relies on a combination of genetic and clinical features. The

discussion then moved to the current management of SDS (Session II). Besides the hallmark pancreatic insufficiency and neutropenia, SDS patients have an increased risk of acute myeloid leukemia (AML) and neurodevelopmental problems, which can include attention deficit hyperactivity disorder (ADHD) and autism spectrum disorders. Session III then proceeded with a panel discussion and workshop covering six international registries of SDS patients. Common problems encountered in establishing and maintaining these registries include difficulty in standardizing patient information from different physicians, designing robust databases, and defining inclusion criteria. The first day concluded with a session on genetic models of SDS (Session IV). Because the *SBDS* gene is highly conserved, informative models have been created in organisms as diverse as fruit flies, mice, and zebrafish.

Day two of the congress was devoted to the function of *SBDS* (Session V). The gene product is a protein that appears to be essential for normal ribosome maturation, suggesting that SDS may be considered an example of a human "ribosomapathy." Although this line of investigation is supported by much compelling data, there is also a possibility that SBDS may have other extraribosomal functions as well. A subsequent session included three talks about *SBDS*'s specific function in hematopoietic maturation,

doi: 10.1111/j.1749-6632.2011.06348.x

 Ann. N.Y. Acad. Sci. 1242 (2011) 26–39 © 2011 New York Academy of Sciences.

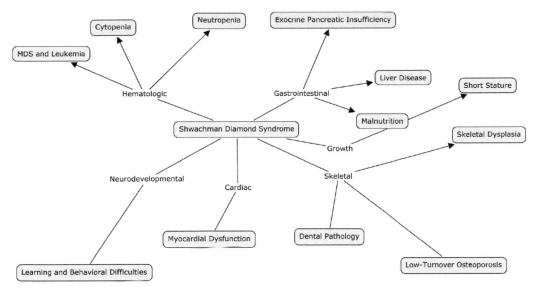

Figure 1. Concept map showing the hematologic, gastrointestinal, growth, skeletal, cardiac, and neurodevelopmental pathology seen in Shwachman-Diamond syndrome. Figure was drawn using IHMC CmapTools software.

introducing the possible role of the stem cell niche in the development of leukemia (Session VI). Day two concluded with a set of presentations about organ development and failure (Session VII), focusing on the neural, pancreatic, and cardiac problems that occur in SDS.

Day three of the congress continued on the pathology of skeletal development in SDS. The final session included two presentations about promising diagnostic and therapeutic technologies (Session VIII). Both of the concluding presentations suggested new possibilities in targeting SDS pathology at the molecular level, which offers hope for the future treatment and prevention of this disease.

Defining the clinical spectrum

Peter Durie (Hospital for Sick Children, Toronto, Canada) opened the session by discussing the clinical spectrum of exocrine pancreatic dysfunction and pointing to its nearly universal manifestation in SDS. The SDS exocrine pancreas shows normal ductular architecture and islets, absent or sparse acinar cells, and extensive fatty replacement. Cross-sectional imaging may reveal a small shrunken pancreas or pancreatic enlargement due to lipomatosis. Furthermore, SDS is a syndrome primarily affecting acinar cells, as analysis of hormonally stimulated pancreatic function studies reveals absent or deficient enzyme secretion but preserved ductal func-

tion, as reflected by fluid and electrolytes. Approximately 50% of SDS patients show moderate improvement in pancreatic acinar capacity, enabling normal digestion of fat and protein without the need for enzyme replacement therapy. Those who become "pancreatic sufficient" with advancing age, show an increase in serum trypsinogen levels. In contrast, the values remain low in those who remain "pancreatic insufficient." In contrast to these age-related changes in serum trypsinogen, all SDS patients have low serum pancreatic isoamylase activities irrespective of age. Thus, changes in the levels of the two enzymes are valuable as a clinical marker of exocrine pancreatic function and as diagnostic markers of the SDS pancreatic phenotype.[5]

Taco Kuijpers (Emma Children's Hospital and Academic Medical Center, Amsterdam, Netherlands) continued, highlighting a set of recently drafted working guidelines for diagnosis and treatment of SDS. Kuijpers also presented two cases that his team diagnosed with SDS, even though they did not display all of the typical traits of the condition.

Sanna Toiviainen-Salo (Helsinki University Hospital, Finland) discussed the use of different imaging modalities in the assessment of pancreatic, hepatic, cardiac, skeletal, and brain development and function. The advanced applications of conventional imaging modalities, such as X-rays, ultrasound, magnetic resonance imaging (MRI), and bone densitometry have provided versatile tools to assess the characteristics of various organs involved in SDS. Newer applications of MRI have highlighted previously unknown pathology in the SDS brain and heart. MRI of brain has revealed significant reductions in volume, as well as in gray and white matter.[6] Cardiac MRI has revealed subtle defects such as diastolic dysfunction at rest and depressed left ventricular contractility during exercise.[7] In addition, abdominal MRI can now be used to identify a lipomatous pancreas with ductular enhancement, as well as hepatic microcysts.[8] Newer functional modalities include MR elastography, which can document elastic changes in the liver, and PET/SPECT, which can enable assessments of actual tissue function and biochemistry.

Akiko Shimamura (Fred Hutchinson Cancer Research Center, Seattle, Washington) discussed her personal approach to treating hematologic complications of SDS. She pointed out several areas of controversy, including the definition of myelodysplastic syndrome (MDS), given that criteria established by the World Health Organization (WHO) are difficult to apply to bone marrow failure patients. Generally, Shimamura reserves the diagnosis of MDS for patients deemed to be at high risk for malignant transformation. Data are also sparse to guide clinical management of acute myeloid leukemia (AML). The role of pretransplant cytoreductive chemotherapy for AML remains to be clarified. Indications for hematopoietic stem cell transplantation include severe or symptomatic cytopenias, MDS, and AML. Encouraging results with reduced intensity conditioning regimens have been reported but patient numbers are still small with limited follow up. Thus, Shimamura does not consider preemptive hematopoietic stem cell transplantation to be the current standard of care. However, since treatment outcomes are superior if initiated prior to the development of leukemia or complications from severe marrow failure, regular monitoring of blood counts and bone marrow exams with cytogenetics are recommended.

Monica Bessler (Children's Hospital of Philadelphia, University of Pennsylvania) next described a unique model of transition of health care from pediatrician to adult care provider. Existing care guidelines reflect our knowledge of this disease during childhood; however, an increasing number of individuals are surviving to adulthood. Much less is known about the natural history and care needs of these individuals as adults. As a result, both the adult health care system and informal care providers are generally ill prepared to care for these individuals as adults. Furthermore, the transition to adulthood poses many general- and disease- specific challenges.

Elizabeth Kerr (Hospital for Sick Children, Toronto, Canada) discussed a relatively unrecognized aspect of SDS: neuropsychological impairment. Specifically, a spectrum of intellectual functioning, with an overall downward shift from the norm, as well as weaknesses in visual processing, attention, and executive functioning have previously been documented. A large proportion of individuals with *SBDS* mutations also manifest neuropsychiatric problems.[9] A recent objective has been to assess functioning longitudinally. The documented difficulties are believed to be primary consequences of SBDS dysfunction in the brain. Preliminary neuroimaging studies indicate that SDS is associated with abnormalities in white matter tracts.[6] Future studies are needed to understand the degree to which these white matter abnormalities relate to cognitive performance in SDS and to investigate the relationship between white matter integrity and SDS-specific clinical variables.

International SDS patient registries: what do they tell us?

The meeting's third session featured a panel discussion and workshop on SDS and related neutropenia registries. Because patients with this rare condition tend to be scattered widely, registries are essential for collecting and centralizing observational data about the disease. The major questions relate to the role of registries, what data should be collected and whether patient registries should be linked to biorepositories. The purpose of the session was to help investigators understand the mission-defined (what is the purpose?), hypothesis-driven (what is the research question?) patient registry. Furthermore in a time of limited resources, the question of what can be accomplished becomes an important issue. There were no real conclusions but an understanding of the extant and nascent registries, their purpose, achievements, and limitations.

Akiko Shimamura (Fred Hutchinson Cancer Research Center, Seattle, Washington) began the session by describing the North American SDS Registry, which opened in December 2008. Seeking to understand the natural history of the disease and improve diagnosis and treatment, this registry initially set out to collect all of the clinical information on each of its participants. Limited time, money, and supplies have since forced the team to focus mainly on hematological complications. This registry promises a very granular look at the hematological issues facing patients with SDS. Blanche Alter (National Cancer Institute, Bethesda, Maryland) discussed her group's SDS and Bone Marrow Failure Registry, which covers all bone marrow failure syndromes. After presenting data on cancer rates in various populations within the registry, Alter offered some advice: "Patients must be classified properly, because if you have people in your registry, who in retrospect don't have the right disease, it makes it very hard to interpret the data." This particular point was not lost on the discussants who understand that care needs to be taken to define the disease carefully, but not too narrowly, so as to create a true case ascertainment bias. Jean Donadieu (Trousseau Hospital, Paris, France) described his group's registry, which collects patient data from all over France to monitor leukemia risk in SDS and a few other disorders. Yigal Dror (The Hospital for Sick Children, University of Toronto, Canada) talked about the Canadian Inherited Marrow Failure Registry, which aims to understand the clinical phenotype of SDS and other bone marrow failure syndromes. This is a very broad registry that can answer some very broad questions but by design does not have tight control over disease definitions. Daniela Longoni (Clinica Pediatrica-Ospedale San Gerardo-Monza, Italy) summarized the data from another European registry. Longoni and her colleagues get medical records as well as bone marrow and blood samples from SDS patients seen in clinics scattered all over Italy. This registry has strived to link the biorepository to granular clinical data and reflects cohesion in the medical community around SDS that may be difficult to attain in North America and, in particular, the United States. Cornelia Zeidler (Hanover Medical School, Hanover, Germany) offered a perspective on the situation in Germany. Since 1994, her group has been collecting information on patients with rare disorders that lead to neutropenia, with a special focus on secondary events such as leukemia and osteoporosis.

In the workshop discussion, the group seemed to reach a consensus that while no registry design would ever be perfect, these efforts provide critical data for understanding the complex pathogenesis of SDS. Furthermore it may take decades to realize the fruits of a registry. Patience, collegiality, and financial support are critical to the success of these endeavors.

Genetics, disease models, and p53

Johanna Rommens (Hospital for Sick Children, Toronto, Canada) opened the session on Genetics and Disease Models by recapitulating genetic studies on *SBDS*, which was identified by positional cloning.[4] Common recurring mutations that lead to premature truncation or aberrant splicing have arisen by gene conversion events to comprise the majority of all patient mutations (76%), with over forty additional rare mutations also reported to date. It is evident that complete loss of SBDS function is not compatible with life, as no patient has been described with two null alleles. The possibility that additional genes may lead to SDS remains an open question, and although introduction of molecular genetic testing of *SBDS* has complemented clinical assessment, improved and more discriminating disease definitions continue to be needed. Mouse models with null, disease-associated, and conditional *Sbds* alleles have revealed a general theme of hypocellularity and growth deficiency that is broadly consistent with a ribosomal deficiency, as first suggested from studies of the yeast ortholog, *Sdo1*. Rommens discussed a mouse model with the disease-associated R126T missense mutation. These mice die at birth but their embryological development can be studied up until birth. Major manifestations include decreased myeloid hematopoietic progenitor growth, a delay in skeletal development (ossification), and apoptosis of postmitotic neurons. Genetic ablation of the tumor suppressor p53 rescues premature brain cell death and hematopoietic progenitor growth but not the somatic growth defect of whole animals.

Shengjiang Tan (University of Cambridge, Cambridge, United Kingdom) continued on the theme of genetic models of SDS by showing that *Drosophila Sbds* (*dSbds*) is essential for larval growth and development. Tan introduced the recent finding from Alan Warren's laboratory that the SBDS protein cooperates with the GTPase elongation factor-like 1 (EFL1) to catalyze a conserved late cytoplasmic step in the maturation of nascent 60S ribosomal subunits.[10] Specifically, SBDS and EFL1 jointly evict the anti-association factor eIF6 from the intersubunit interface of pre-60S ribosomal subunits to allow ribosomal subunit joining. Tan's work suggests that ectopic expression of wild-type eIF6 enhances the growth defect of hypomorphic *dSbds* mutant flies by exacerbating the defect in ribosomal subunit joining. By contrast, transgenic overexpression of dominant gain-of-function eIF6 suppressor mutants completely rescues the lethal phenotype caused by *dSbds* deficiency. Finally, Tan demonstrated that genetic ablation of p53 rescues the cell-cycle arrest, apoptosis, and developmental abnormalities associated with loss of *dSbds* function. Tan hypothesizes a ribosomal stress surveillance pathway in *Drosophila* that activates p53 in an Mdm2-independent manner.

Elayne Provost (Johns Hopkins University, Baltimore, Maryland) used morpholino knockdown of the zebrafish ortholog of S*bds* to recapitulate the disease phenotype: loss of neutrophils, a small exocrine pancreas, and a disrupted skeletal architecture. To assess whether these phenotypes were due to a p53-dependent mechanism, Provost knocked down p53 by co-injection of anti-p53 morpholino and by genetic crosses with a *p53*-null zebrafish line. In both experiments, the SDS phenotype was not rescued. Provost therefore concluded that the SDS zebrafish model was not p53-dependent, in contrast to the two previous models in the session.

Ribosomal and extra-ribosomal functions of SBDS

Steven Ellis (University of Louisville, Louisville, Kentucky) opened the session on SBDS function by revisiting the slow growth phenotype associated with the deletion of *SDO1*, the yeast ortholog of *SBDS*. This slow growth phenotype is associated with defects in the maturation of 60S ribosomal subunits and can be suppressed by mutations in the *TIF6* gene, suggesting a role for Sdo1 in recycling Tif6 from 60S subunits in the cytoplasm back to the nucleus.[11] He then reported that yeast cells lacking Sdo1 fail to grow on media containing only respiratory carbon sources, suggestive of a defect in mitochondrial energy metabolism. Related studies in human TF-1 cells revealed that cells depleted of SBDS also exhibit reduced oxygen consumption relative to controls. Given that the largest producer of reactive oxygen species is the mitochondrial elec-

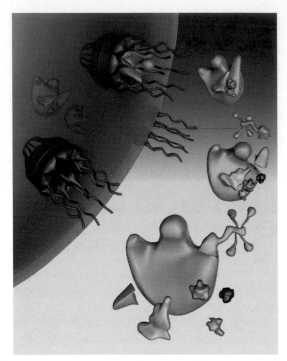

Figure 2. In eukaryotic cells, ribosomes must be shipped from their site of assembly in the nucleus (represented in purple) to the cytoplasm (represented in grayish blue), where they function in mRNA translation and protein synthesis. Surprisingly, newly made ribosomes do not arrive at their new job ready to get to work. Instead, they arrive in a functionally inactive form and require some "on-site" unpacking and assembly. Unpacking involves the removal of the small entourage of transacting factors that facilitate ribosome assembly and export. Some of these factors also prevent the nascent subunits from engaging prematurely with components of the translation machinery. Yet others act as placeholders for the few ribosomal proteins that are added in the cytoplasm. One critical event for the large (60S) subunit is assembly of the stalk (shown in the figure), a structure that is required for recruitment and activation of the GTPases of translation. The figure was drawn by Marianna Grenadier © and provided by Dr. Arlen Johnson, University of Texas.

tron transport chain, perturbation of respiratory function in cells depleted of Sdo1 or SBDS could be a potential source of elevated reactive oxygen species (ROS). To investigate the potential molecular mechanisms underlying these respiratory deficient phenotypes, Ellis carried out a proteomic analysis comparing yeast cells depleted of Sdo1 with controls. His data revealed that cells lacking Sdo1 overexpress Por1, an ortholog of human VDAC1, a voltage-dependent anion channel believed to be an essential component of the mitochondrial permeability pore. Both over- and underexpression of

VDAC1 has been shown to disrupt mitochondrial function and lead to apoptosis. Ellis speculates that SBDS may be a multifunctional protein affecting processes other than ribosome synthesis.

Arlen Johnson (University of Texas, Austin, Texas) began the first of five presentations focusing on the role of translation initiation factor 6 (Tif6) or (eukaryotic initiation factor 6 (eIF6) in mammals) in ribosome biogenesis. Johnson focused on the maturation process by which eukaryotic ribosomes are preassembled in the nucleus and mature in the cytoplasm (Fig. 2). The nascent subunits entering the cytoplasm are functionally inactive because of the presence of factors that preclude their activity (Tif6) as well as the absence of ribosomal proteins (Fig. 2). One critical event for the large (60S) subunit is assembly of the stalk, a structure that is required for recruitment and activation of the GTPases of translation. During maturation, assembly of the stalk is a prerequisite for the release of the anti-association factor Tif6 by the translocase-like GTPase Efl1. Johnson reported that an internal loop of the large ribosomal protein Rpl10, which embraces the P site tRNA, is also required for release of Tif6, only 90Å away. Mutations in this P site loop blocked 60S maturation but were suppressed by mutations in Tif6 or Efl1. The Efl1 mutations mapped to domain interfaces important for conformational changes of EF-G and eEF2 during translocation. Molecular dynamics simulations of the mutant Efl1 proteins predict that they promote a conformation in Efl1 equivalent to the translocational intermediate of EF-G. Johnson hypothesizes molecular signaling from the ribosomal P site to Tif6, via Efl1, suggesting that the integrity of the P site is interrogated during maturation of the 60S subunit. Johnson further proposes that Efl1 undergoes a conformational change that, analogous to eEF2 during translocation, serves as a quasi-functional check of the integrity of the 60S subunit prior to its first round of *bona fide* translation.

Alan Warren (University of Cambridge, Cambridge, United Kingdom) expanded the discussion of 60S ribosomal subunit maturation by highlighting the critical role of removal of the assembly factor eIF6, which imposes a physical barrier to ribosomal subunit joining. First, Warren recapitulated genetic studies in yeast that indicate a requirement for the ortholog of the *SBDS* gene, *Sdo1*.[12] He then described experiments using mice with a

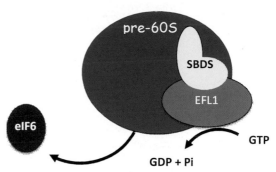

Figure 3. Model of eIF6 release by SBDS and EFL1. SBDS stimulates 60S-dependent GTP hydrolysis by EFL1, generating EFL1.GDP.Pi. Following release of inorganic Pi, EFL1 adopts its GDP-bound conformation and domain I of SBDS is rotated relative to domains II and III, directly or indirectly disrupting the intersubunit bridge B6. Binding of eIF6 is destabilized, release of eIF6 is triggered, and EFL1.GDP and SBDS dissociate from the ribosome. Release of eIF6 allows the formation of actively translating 80S ribosomes. The figure is courtesy of Dr. Alan Warren, University of Cambridge.

conditional deletion of *Sbds*. By isolating late cytoplasmic 60S ribosomal subunits from these *Sbds*-deleted mice, Warren showed that Sbds and the GTPase elongation factor-like 1 (Efl1) directly catalyze eIF6 removal in mammalian cells by a mechanism that requires GTP binding and hydrolysis by Efl1[10] (Fig. 3). Warren then analyzed disease-associated missense variants to suggest that the essential role of SBDS is to tightly couple GTP hydrolysis by EFL1 on the ribosome to eIF6 release. Complementary NMR structural and dynamic studies of the human and yeast SBDS proteins suggest mechanistic parallels between this late step in 60S maturation and aspects of bacterial ribosome disassembly. In order to prove the generality of his model, Warren turned to experiments using Dictyostelium,[13] which largely confirmed the cooperative role of SBDS and EFL1 in the release of eIF6. Warren concluded by emphasizing the direct role for SBDS and EFL1 in catalyzing the translational activation of ribosomes in all eukaryotes and defining SDS as a ribosomopathy caused by uncoupling GTP hydrolysis from eIF6 release.

Johnson Liu (The Feinstein Institute for Medical Research, Manhasset, New York) continued on the eIF6 theme, discussing experiments in human TF-1 erythroleukemia and A549 carcinoma cells with partial or near-complete silencing of SBDS by RNA interference. The growth and hematopoietic colony

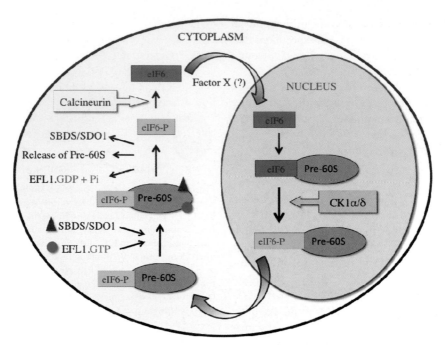

Figure 4. Schematic representation of the functional pathway of eIF6 in 60S ribosome biogenesis. In the nucleolus, eIF6 associates with the pre-60S particles along with >100 transacting protein factors and is essential for pre-60S ribosome assembly and pre-rRNA processing. eIF6 remains associated with the pre-60S particles during pre-60S maturation in the nucleoplasm as well as during the subsequent nuclear export of the pre-60S particles. Nuclear export of eIF6 bound to the pre-60S particles requires phosphorylation of eIF6 at Ser-174 and Ser-175 by the nuclear isoform of CK1. In the cytoplasm, during the final maturation process, two-cytoplasmic proteins SBDS/Sdo1 and EFL1/Efl1p interact with the pre-60S particles and catalyze the release of eIF6 coupled to GTP hydrolysis by EFL1. The released eIF6 that is presumably in the phosphorylated form, then interacts with Ca^{2+}/calmodulin-regulated protein phosphatase calcineurin and the dephosphorylated form of eIF6, either by itself, or by interaction with another as yet unidentified protein factor X containing the NLS signal, is imported to the nucleolus to participate in another round of 60S ribosome biogenesis. The figure is courtesy of Dr. Umadas Maitra, Albert Einstein College of Medicine of Yeshiva University.

forming potential of TF-1 knockdown cells were markedly hindered compared to controls. To understand the effect of SBDS on 60S subunit maturation, subunit localization was assessed in SBDS-depleted A549 cells by transfection with a vector expressing a fusion between the large ribosomal subunit protein RPL29 and green fluorescence protein. These experiments suggested a higher percentage of SBDS-depleted cells with nuclear localization of 60S subunits. Liu also analyzed the levels of eIF6 following near-complete knockdown of SBDS in TF-1 cells. The percentage of eIF6 associated with 60S subunits increased 1.5-fold in the SBDS knockdown samples. Liu concluded that knockdown of SBDS leads to growth inhibition and defects in ribosome maturation, suggesting a role for wild-type SBDS in nuclear export of pre-60S subunits and eIF6 recycling.

Umadas Maitra (Albert Einstein College of Medicine of Yeshiva University, Bronx, New York)

expanded on the history of research on eIF6, a highly conserved protein from yeast to mammals, essential for 60S ribosome biogenesis and assembly, which is mostly a nucleolar function. Association of eIF6 with the pre-60S ribosomal particles is also required for the export of the pre-60S particles from the nucleus to the cytoplasm where the release of eIF6 occurs. Both yeast and mammalian eIF6 are phosphorylated at Ser-174 and Ser-175 by the nuclear isoforms of casein kinase 1 (CK1). However, the molecular basis of eIF6 phosphorylation remains elusive. Maitra has shown that eIF6 shuttles continuously between the nucleus and the cytoplasm, and this shuttling is mediated by the opposing action of CK1 and the Ca^{2+}/calmodulin-dependent protein phosphatase, calcineurin (Fig. 4). Maitra demonstrated that calcineurin, following its activation by Ca^{2+}, binds to and promotes rapid translocation of eIF6 from the cytoplasm to the nucleus, an event that is blocked by specific calcineurin inhibitors cyclosporin A or FK520, and suggesting that the dephosphorylated form of eIF6 is imported to the nucleus. The nuclear export of eIF6, on the other hand, requires rephosphorylation of Ser-174 and Ser-175. Failure to phosphorylate at these sites either by mutation of the serine residues to alanine or treatment of cells with a specific CK1 inhibitor inhibits nuclear export of eIF6 and results in nuclear accumulation of eIF6. Together, these results establish eIF6 as a substrate for calcineurin and suggest a novel paradigm of calcineurin function in 60S ribosome biogenesis via regulating the nuclear accumulation of eIF6.

Stefano Biffo (San Raffaele Scientific Institute, Milan, Italy) concluded the five talks focusing on eIF6. Biffo began by recapitulating the phenotype of mice that are heterozygous for *eIF6*. These animals have a subtle growth phenotype involving fat cell and liver abnormalities, reduced cytoplasmic eIF6, delayed G1/S cell cycle progression, reduced translation, and an impaired response to insulin stimulation. Antagonizing eIF6 activity also has a protective effect on lymphomagenesis and tumor growth. Conversely, restoring eIF6 activity recovers insulin sensitivity but increases oncogene-induced transformation. Biffo then discussed the possibility that eIF6 agonists/antagonists may be therapeutically useful in SDS. To progress in the generation of active modulators of eIF6 binding to 60S, Biffo developed a microwell-based screening assay to be employed for the identification of small molecules that affect the binding of eIF6 to 60S subunits. Biffo concluded by proposing to test whether antagonizing eIF6 would be beneficial in SDS by crossing *eIF6* heterozygous mice with *Sbds* mutant mice.

Akiko Shimamura reported on both ribosomal and extraribosomal functions of SBDS. Similar to observations in yeast and mouse systems, she found that the 60S:40S ribosomal subunit ratio is consistently reduced in cells from SDS patients and that patient cells exhibit a reduced capacity to form 80S subunits from 40S and 60S subunits *in vitro*. Addition of wild-type SBDS or depletion of eIF6 improved ribosome joining. She then discussed the putative function of SBDS in mitotic spindle stabilization. Previously, Shimamura showed that SBDS directly binds and stabilizes microtubules in vitro. SBDS loss *in vivo* resulted in sensitivity to nocodazole and resistance to taxol. Total internal reflection fluorescence microscopy studies demonstrate that SBDS binds to purified microtubules with a K_d similar to that of other microtubule-associated proteins. SBDS loss results in shortened microtubule length and decreased microtubule acetylation. SBDS depletion of human CD34$^+$ cells results in marked reduction of hematopoietic progenitor colony formation, and hematopoiesis is improved by the addition of taxol, a microtubule stabilizer.

André Valentina (Universita di Milano-Biccoca, Milan, Italy) concluded this session by presenting preliminary analysis of the functional properties of bone marrow (BM)-derived mesenchymal stem cells (MSCs). MSC were obtained from 27 SDS patients. At the third passage of the culture, MSC were tested for the expression of specific surface markers, their ability to differentiate into mesengenic lineages, their capability to abrogate T cell proliferation, and their capacity to prevent neutrophil apoptosis. MSCs derived from SDS patients (SDS-MSCs) displayed typical fibroblastoid morphology and expressed common MSC markers. These cells were able to differentiate into adipocytes and osteoblasts. In addition, SDS-MSCs drastically decreased mitogen-induced lymphocyte proliferation. SDS-MSCs were comparable to healthy donor (HD)-MSCs in supporting the viability of neutrophils. SDS-MSCs were also able to produce high amounts of IL-6, a cytokine involved in the protection of neutrophils from apoptosis. Genome-wide gene expression analysis was carried out and preliminary results showed a specific profile for SDS-MSCs, which was significantly different from HD-MSCs.

SBDS in hematopoiesis and leukemia: role of the stem cell niche

Paul Frenette (Albert Einstein College of Medicine of Yeshiva University, Bronx, New York) opened the session on SBDS in Hematopoiesis and Leukemia by discussing new concepts on the role of the stem cell niche (Fig. 5). The existence of a stem cell niche has been postulated but the identity of the mesenchymal stem cell has been elusive until recently. Frenette discussed his work on nestin$^+$ cells that express Cxcl12 (also known as Sdf-1).[14] These cells were discovered in the course of experiments on circadian fluctuations in hematopoietic stem cell (HSC) homing, which showed efficient homing at night and ablation by surgical sympathectomy. Conversely, β3 adrenergic receptor agonists enhanced homing. Frenette

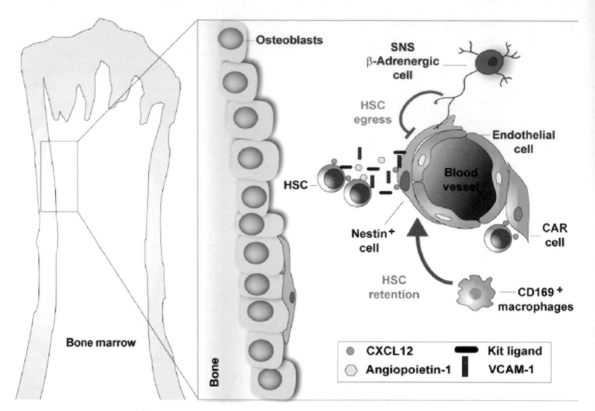

Figure 5. Mesenchymal niche for hematopoiesis indicating egress and retention of hematopoietic stem cells (HSC). The figure is courtesy of Dr. Paul Frenette, Albert Einstein College of Medicine of Yeshiva University, and modified from Ref. 32.

confirmed that nestin$^+$ cells constituted part of a neuro-reticular complex, in which nestin$^+$ cells are targeted by the sympathetic nervous system, are adjacent to HSC, and express Cxcl12. Nestin$^+$ cells form self-renewing clonal spheres and contribute to bone formation. Finally, Frenette concluded by discussing a second pathway whereby the stem cell niche modulates trafficking of HSC. Depletion of mononuclear phagocytes (CD169$^+$ macrophages) can induce egress of HSC[15] (Fig. 5).

David Scadden (Massachusetts General Hospital, Harvard Medical School, Boston, Massachusetts) delivered the Jason Bennette Memorial Lecture following a brief introduction by Jeffrey Lipton. Scadden proposed that mesenchymal–parenchymal interactions are critical in development and can participate in adult tissue homeostasis as exemplified by bone–bone marrow interactions in hematopoiesis, particularly the hematopoietic stem cell (HSC) niche. First, Scadden reviewed experiments in which a key gene product involved in microRNA production, Dicer1, was specifically deleted in osteoprogen-

itors of the bone mesenchymal lineage.[16] Scadden demonstrated that these specific genetic alterations could induce complex secondary changes in the organization of the hematopoietic (parenchymal) lineage, including the development of independent genetic mutations and frank leukemia (Fig. 6). This model supports the hypothesis that mesenchymal cells comprising tissue stroma may serve as the initiating "hit" in the multi-hit process of oncogenesis. Scadden then discovered that conditional deletion of Sbds in osteoprogenitors also resulted in a myelodysplastic phenotype.[16] Finally, Scadden examined the dynamics of the mesenchymal population in bone marrow, in order to assess the potential for acquired genetic lesions in the niche contributing to hematopoietic disease. For these experiments, he focused on mesenchymal stem cells (MSC) labeled by Mx-1. Using genetic pulse-chase experiments, Scadden found that mesenchymal cells have an unexpectedly high turnover rate, are dependent upon a stem/progenitor population for replenishment, and can transit in solid tissue. Therefore, the

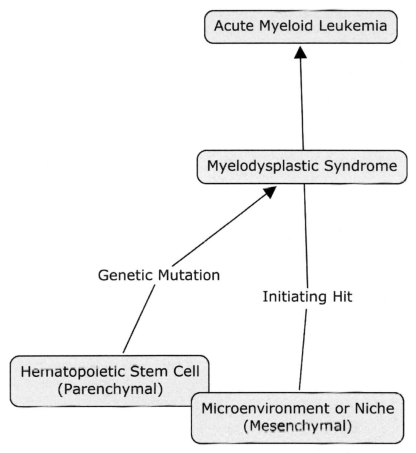

Figure 6. Concept map showing that mesenchymal–parenchymal interactions are critical in development and can participate in bone–bone marrow interactions modulating the hematopoietic stem cell (HSC) niche and leading to myelodysplastic syndrome and acute myeloid leukemia. Figure was drawn using IHMC CmapTools software.

mesenchymal cells of tissue "stroma" may be a highly dynamic population that can play a central role in normal and malignant tissue biology.

Yigal Dror summarized hematopoietic pathology in SDS, citing both his own and others' published findings. Dror cited studies suggesting that SBDS is critical for hematopoietic cell expansion and colony formation.[17,18] Dror also found accelerated apoptosis and oxidative stress in SBDS-deficient cells undergoing erythroid differentiation. During erythroid differentiation of K562 cells, Dror further found that the ribosome profile and global translation were more severe than in resting cells. Dror also pointed out the high propensity in SDS for clonal and malignant myeloid transformation including myelodysplastic syndrome (MDS) and acute myeloid leukemia (AML). The most common clonal marrow cytogenetic abnormalities in SDS are i(7q) and del(12)(q12). However, both may not be associated with malignant progression and even may regress to levels below detection. This is in contrast to monosomy of chromosome 7 or part of the long arm of chromosome 7. The precise molecular mechanism of leukemia is unknown. Dror suggested that mitotic spindle instability may increase the risk of chromosome abnormalities. Ribosome defects, slow cell expansion and apoptosis may decrease the ability of nonmalignant cells to compete and control the expansion of malignant clones. Increased angiogenesis and decreased immunity may further enable progression of malignant clones.

Brain, pancreas, heart, and bone: development and failure

In the session entitled "Organ Development and Failure," Russell Schachar (University of Toronto,

Toronto, Canada) discussed the latest developments in our understanding and treatment of attention deficit hyperactivity disorder (ADHD) and its implications for managing the neurocognitive aspects of SDS. ADHD[19–22] is a highly heritable disorder characterized by developmentally atypical and impairing restlessness, impulsiveness, and inattentiveness, which commences in early childhood and persists, in many cases, throughout life. ADHD responds well to stimulants such as methylphenidate and dextroamphetamine. ADHD is diagnosed when other psychiatric and neurological disorders are absent. Nevertheless, ADHD often occurs along with conditions such as autism, and is not uncommon in SDS. There seems to be genetic overlap among these various conditions and neuropathological overlap with ADHD that is secondary to brain injury. Yet, "ADHD" is typically overlooked when these other conditions, such as SDS, are evident. Consequently, these children may not receive treatment for their ADHD. Although further research is required before drawing any firm conclusions, Schachar suggests that treatment for SDS could be improved by incorporating some of the lessons learned about ADHD.

In keeping with this theme of understanding the genesis of organ dysfunction in SDS, Sabrina Desbordes (Center for Genomic Regulation, Barcelona, Spain), described depletion of the Sbds protein in different neural and neural progenitor cells derived from mouse and human embryonic stem (ES) cells. Since the brain is affected in SDS, development of novel techniques to study neural cell function is important. To this end, Desbordes and her colleagues developed coculture protocols for the derivation of neural, neuronal, and glial cell types from mouse and human ES cells, including neural stem cells, motor neurons, dopaminergic neurons, serotonin neurons, GABA neurons, astrocytes, and oligodendrocytes. Desbordes also has pioneered the development of phenotypic cell-based assays for drug discovery using high-throughput screening in human ES cells. Using a depletion system, Desbordes studied the role of SBDS during neuronal cell type specification and differentiation of glial cells. She plans to conduct screens to identify compounds with a positive effect on SBDS-depleted neurons.

Steven Leach (McKusick-Nathans Institute of Genetic Medicine, Johns Hopkins University, Baltimore, Maryland) reported on the field of pancreas development in mice and zebrafish. Formation of the vertebrate pancreas requires a highly coordinated series of developmental events, including anterior–posterior and ventral–dorsal foregut patterning, branching morphogenesis, and the progressive differentiation of endocrine and exocrine cells from multilineage progenitors. In the mouse, dorsal and ventral domains of posterior foregut endoderm form epithelial buds, which grow and eventually resolve into branched epithelial trees containing both endocrine and exocrine progenitor cells. In zebrafish, similar morphogenetic events can be identified, with the added advantage of real-time examination in living embryos. In both mouse and zebrafish, the subsequent differentiation of endocrine and exocrine cell types is guided by a hierarchical cascade of transcription factors, under tight regulation by the Notch signaling pathway.[23,24] Hence, it was concluded that the mechanisms of pancreatic development are highly conserved and therefore, either of these systems will prove useful in understanding pancreatic pathogenesis in SDS.

Continuing on this theme, Marina Tourlakis and colleagues (Rommens Laboratory, The Hospital for Sick Children, Toronto, Canada) generated mice with a floxed or conditional (CKO) *Sbds* allele, to specifically remove Sbds from pancreatic lineages, in combination with null (KO) and missense (*R126T*) alleles. Loss of Sbds in the pancreas resulted in notably small pancreases, displaying the hallmarks of SDS, with hypoplasia of the exocrine pancreas along with fat infiltration by the time of weaning. Furthermore, these targeted models demonstrate a genotype–phenotype correlation, with CKO/KO models displaying pathology at earlier time points than CKO/R126T models. Acinar cells express amylase; however serum enzyme levels for amylase and lipase were significantly decreased compared with littermate controls. Further, consistent with most patient findings, the SDS pancreas-model mice did not display clinical features of endocrine dysfunction, although decreased islet size in the CKO/KO model indicate requirements for Sbds in endocrine growth. Overall growth of the mice was impaired as early as the perinatal period with body mass and length of mutants smaller than control littermates, pointing to major consequences of nutritional impairment. The targeting of Sbds in the pancreas is sufficient to recapitulate the full SDS pancreatic phenotype. In a follow-up session, the same group presented their findings on acinar pathology in these mice. A mouse model with conditional targeted knockout (*CKO*) and disease-associated missense (*R126T*) alleles was used to study the consequences of *Sbds* mutation. Excision of the *CKO* allele specifically in the pancreas was achieved by Cre recombinase driven by the *Ptf1a* transcription factor promoter. The $Sbds^{CKO/R126T}Ptf1a^{Cre/+}$ pancreas displays hypoplasia and progression to severe dysplasia by young adulthood. Acini appear disorganized and small, containing cells with reduced amylase staining that were not apoptotic. Interestingly, increased staining for markers of proliferation such as PCNA and Ki67 was evident. The β-catenin and c-Myc members of the Wnt signaling pathway, known to mediate acinar cell expansion, were also increased. The timing of the severe dysplasia coincides with weaning and the most rapid growth period known for exocrine tissue. Furthermore, acinar cells are highly specialized for the production, storage, and secretion of digestive enzymes. Tourlakis suggested that these unique growth demands and the specialized exocrine proteome sensitize the pancreas to translational impairment caused by mutations in *Sbds*.

Ibrahim Domian (Massachusetts General Hospital, Harvard Medical School, Boston, Massachusetts) next discussed development of the heart, which exhibits subtle abnormalities in SDS. Recent studies have suggested a small reduction in cardiac contractility, which only becomes significant when patients are stressed. To understand how the heart develops in healthy and diseased individuals, Domian has developed methods for differentiation of mouse embryonic stem cells to cardiomyocytes, and he is using a high-throughput approach to identify compounds that enable expansion of these cells.

The other major organ that affected in SDS is the skeleton (bone). Outi Makitie (Children's Hospital, Helsinki University Central Hospital and University of Helsinki, Finland) reviewed the current understanding of the group of disorders classified as "metaphyseal chondrodysplasias," which includes SDS. SDS bone dysplasia is characterized by short stature, delayed appearance but subsequent normal development of secondary ossification centers, and by variable metaphyseal widening and irregularity most often seen in the ribs in early childhood and in the proximal and distal femora later in childhood and adolescence.[25] Rarely, skeletal in-

volvement may be extremely severe with generalized bone abnormalities. Although metaphyseal changes often become undetectable and clinically insignificant over time, they may also progress and result in limb deformities at the hips and the knees, or stress fractures of the femoral necks. SDS bone disease also includes early-onset low-turnover osteoporosis, which is characterized by low bone mass and vertebral fragility fractures.[26] Histological and histomorphometric analyses of transiliac bone biopsies show reduced numbers of osteoblasts and osteoclasts and low trabecular bone volume consistent with primary osteoporosis. Low-turnover osteoporosis may result from a primary defect in bone metabolism that is related to the bone marrow dysfunction and neutropenia. In mice, Sbds is required for *in vitro* and *in vivo* osteoclastogenesis.[27] Impaired osteoclast formation may disrupt bone homeostasis and result in the skeletal abnormalities seen in SDS patients. Optimal treatment for SDS osteoporosis remains to be established.

William Cole (University of Alberta, Edmonton, Canada) next expanded upon the general theme of skeletal dysplasias. The individual dysplasias have distinctive clinical, radiographic, and molecular features. Most display a wide range of severities and there is often clinical and molecular overlap between the different types. Cole reviewed most of the genes associated with autosomal dominant disorders of skeletal development.[28] However, genes responsible for autosomal recessive disorders remain elusive. Finally, Gerard Karsenty (Columbia University Medical Center, New York, New York) provided substantial evidence linking energy metabolism to bone mass and thus introducing the concept of bone as an "endocrine organ."[29] Critical in this discussion is the role of leptin and its effects on serotonin metabolism.[30,31]

New targets for therapy in SDS

M. William Lensch (Children's Hospital Boston, Harvard Medical School, Boston, Massachusetts) opened the session on Novel Diagnostics and Therapeutics by describing the creation of human pluripotent stem cell models of SDS by two methods: knock-down of the SBDS gene by lentiviral RNAi in human embryonic stem cells (hESCs), and generation of induced pluripotent stem cell (iPSC) lines from two SDS patients. Both SDS hESCs and iPSCs manifest specific deficits in exocrine pancreatic and hematopoietic differentiation, and these cellular defects can be rescued by SBDS overexpression. Lensch provocatively suggests that protease-mediated auto-digestion contributes to the pancreatic and hematopoietic phenotypes in SDS.

Paul de Figueiredo (Texas A & M University, College Station, Texas) concluded the session by proposing a new biochemical function for SBDS. He first described a large-scale screen for compounds that reverse the disease phenotype of *Saccharomyces cerevisiae* cells depleted of Sdo1, the yeast ortholog of SBDS: histone deacetylase (HDAC) inhibitors were found as top hits able to rescue the slow growth defect. Subsequently, de Figueiredo demonstrated that Sdo1/SBDS inhibited HDAC enzymatic activities in HeLa cell nuclear extracts. HDAC inhibition also promoted the growth of hematopoietic progenitor cells depleted of SBDS by siRNA. It is now known that acetylation can occur on many proteins other than histones, and these observations suggest that dysregulation of acetylation may constitute a component of the SDS pathology.

Conflicts of interest

The authors declare no conflicts of interest.

References

1. Shwachman, H., L.K. Diamond, F.A. Oski & K.T. Khaw. 1964. The syndrome of pancreatic insufficiency and bone marrow dysfunction. *J. Pediatr.* **65:** 645–663.
2. Bodian, M., W. Sheldon & R. Lightwood. 1964. Congenital Hypoplasia of the exocrine pancreas. *Acta. Paediatr.* **53:** 282–293.
3. Goobie, S., M. Popovic, J. Morrison, *et al.* 2001. Shwachman-Diamond syndrome with exocrine pancreatic dysfunction and bone marrow failure maps to the centromeric region of chromosome 7. *Am. J. Hum. Genet.* **68:** 1048–1054.
4. Boocock, G.R., J.A. Morrison, M. Popovic, *et al.* 2003. Mutations in SBDS are associated with Shwachman-Diamond syndrome. *Nat. Genet.* **33:** 97–101.
5. Ip, W.F., A. Dupuis, L. Ellis, *et al.* 2002. Serum pancreatic enzymes define the pancreatic phenotype in patients with Shwachman-Diamond syndrome. *J. Pediatr.* **141:** 259–265.
6. Toiviainen-Salo, S., O. Makitie, M. Mannerkoski, *et al.* 2008. Shwachman-Diamond syndrome is associated with structural brain alterations on MRI. *Am. J. Med. Genet. A.* **146A:** 1558–1564.
7. Toiviainen-Salo, S., O. Pitkanen, M. Holmstrom, *et al.* 2008. Myocardial function in patients with Shwachman-Diamond syndrome: aspects to consider before stem cell transplantation. *Pediatr. Blood Cancer* **51:** 461–467.
8. Toiviainen-Salo, S., M. Raade, P.R. Durie, *et al.* 2008. Magnetic resonance imaging findings of the pancreas in patients with Shwachman-Diamond syndrome and mutations in the SBDS gene. *J. Pediatr.* **152:** 434–436.
9. Kerr, E.N., L. Ellis, A. Dupuis, J.M. Rommens & P.R. Durie. 2010. The behavioral phenotype of school-age children with shwachman diamond syndrome indicates neurocognitive dysfunction with loss of Shwachman-Bodian-Diamond syndrome gene function. *J. Pediatr.* **156:** 433–438.
10. Finch, A.J., C. Hilcenko, N. Basse, *et al.* 2011. Uncoupling of GTP hydrolysis from eIF6 release on the ribosome causes Shwachman-Diamond syndrome. *Genes. Dev.* **25:** 917–929.
11. Moore, J.B., J.E. Farrar, R.J. Arceci, *et al.* 2010. Distinct ribosome maturation defects in yeast models of Diamond-Blackfan anemia and Shwachman-Diamond syndrome. *Haematologica* **95:** 57–64.
12. Menne, T.F., B. Goyenechea, N. Sanchez-Puig, *et al.* 2007. The Shwachman-Bodian-Diamond syndrome protein mediates translational activation of ribosomes in yeast. *Nat. Genet.* **39:** 486–495.
13. Wong, C.C., D. Traynor, N. Basse, *et al.* 2011. Defective ribosome assembly in Shwachman-Diamond syndrome. *Blood* **118:** 4305–4312.
14. Mendez-Ferrer, S., T.V. Michurina, F. Ferraro, *et al.* 2010. Mesenchymal and haematopoietic stem cells form a unique bone marrow niche. *Nature* **466:** 829–834.
15. Chow, A., D. Lucas, A. Hidalgo, *et al.* 2011 Bone marrow CD169+ macrophages promote the retention of hematopoietic stem and progenitor cells in the mesenchymal stem cell niche. *J. Exp. Med.* **208:** 261–271.
16. Raaijmakers, M.H., S. Mukherjee, S. Guo, *et al.* 2010. Bone progenitor dysfunction induces myelodysplasia and secondary leukaemia. *Nature* **464:** 852–857.
17. Yamaguchi, M., K. Fujimura, H. Toga, *et al.* 2007. Shwachman-Diamond syndrome is not necessary for the terminal maturation of neutrophils but is important for maintaining viability of granulocyte precursors. *Exp. Hematol.* **35:** 579–586.
18. Rawls, A.S., A.D. Gregory, J.R. Woloszynek, *et al.* 2007. Lentiviral-mediated RNAi inhibition of Sbds in murine hematopoietic progenitors impairs their hematopoietic potential. *Blood* **110:** 2414–2422.
19. Polanczyk, G., M.S. de Lima, B.L. Horta, *et al.* 2007. The worldwide prevalence of ADHD: a systematic review and metaregression analysis. *Am. J. Psychiatry.* **164:** 942–948.
20. Mental health in the United States. 2005. Prevalence of diagnosis and medication treatment for attention-deficit/hyperactivity disorder–United States, 2003. *MMWR. Morb. Mortal. Wkly. Rep.* **54:** 842–847.
21. Martin, N.C., J.P. Piek, D. Hay. 2006. DCD and ADHD: a genetic study of their shared aetiology. *Hum. Mov. Sci.* **25:** 110–124.
22. Baker, B.L., C.L. Neece, R.M. Fenning, *et al.* 2010. Mental disorders in five-year-old children with or without developmental delay: focus on ADHD. *J. Clin. Child. Adolesc. Psychol.* **39:** 492–505.
23. Lin, J.W., A.V. Biankin, M.E. Horb, *et al.* 2004. Differential requirement for ptf1a in endocrine and exocrine lineages of developing zebrafish pancreas. *Dev. Biol.* **274:** 491–503.

24. Esni, F., B. Ghosh, A.V. Biankin, *et al.* 2004. Notch inhibits Ptf1 function and acinar cell differentiation in developing mouse and zebrafish pancreas. *Development* **131:** 4213–4224.

25. Makitie, O., L. Ellis, P.R. Durie, *et al.* 2004. Skeletal phenotype in patients with Shwachman-Diamond syndrome and mutations in SBDS. *Clin. Genet.* **65:** 101–112.

26. Toiviainen-Salo, S., M.K. Mayranpaa, P.R. Durie, *et al.* 2007. Shwachman-Diamond syndrome is associated with low-turnover osteoporosis. *Bone* **41:** 965–972.

27. Leung, R., K. Cuddy, Y. Wang, *et al.* 2011. Sbds is required for Rac2-mediated monocyte migration and signaling downstream of RANK during osteoclastogenesis. *Blood* **117:** 2044–2053.

28. Makitie, O., M. Susic, L. Ward, *et al.* 2005. Schmid type of metaphyseal chondrodysplasia and COL10A1 mutations—findings in 10 patients. *Am. J. Med. Genet. A.* **137A:** 241–248.

29. Rached, M.T., A. Kode, B.C. Silva, *et al.* 2010. FoxO1 expression in osteoblasts regulates glucose homeostasis through regulation of osteocalcin in mice. *J Clin Invest* **120:** 357–368.

30. Yadav, V.K., F. Oury, N. Suda, *et al.* 2009. A serotonin-dependent mechanism explains the leptin regulation of bone mass, appetite, and energy expenditure. *Cell* **138:** 976–989.

31. Oury, F., & G. Karsenty. 2011. Towards a serotonin-dependent leptin roadmap in the brain. *Trends. Endocrinol. Metab.* **22:** 382–387.

32. Pinho, S., D. Lucas & P.S. Frenette. (?) *Mesenchymal Stromal Cells: Basic Biology & Clinical Applications.* Humana Press. Clifton, New Jersey.

Ann. N.Y. Acad. Sci. ISSN 0077-8923

ANNALS OF THE NEW YORK ACADEMY OF SCIENCES
Issue: Annals *Meeting Reports*

Draft consensus guidelines for diagnosis and treatment of Shwachman-Diamond syndrome

Yigal Dror,[1] Jean Donadieu,[2] Jutta Koglmeier,[3] John Dodge,[4] Sanna Toiviainen-Salo,[5] Outi Makitie,[5] Elizabeth Kerr,[1] Cornelia Zeidler,[6] Akiko Shimamura,[7] Neil Shah,[3] Marco Cipolli,[8] Taco Kuijpers,[9] Peter Durie,[1] Johanna Rommens,[1] Liesbeth Siderius,[10] and Johnson M. Liu[11]

[1]The Hospital For Sick Children, University of Toronto, Ontario, Canada. [2]Trousseau Hospital, Paris, France. [3]Great Ormond Street Hospital and Institute of Child Health, London, UK. [4]University of Wales Swansea, UK. [5]Helsinki University Hospital and Children's Hospital, University of Helsinki, Helsinki, Finland. [6]Hannover Medical School, Hannover, Germany. [7]Fred Hutchinson Cancer Research Center, University of Washington, Seattle, Washington. [8]Cystic Fibrosis Center, Ospedale Civile Maggiore, Verona, Italy. [9]Emma Children's Hospital, Academic Medical Center, University of Amsterdam, the Netherlands. [10]Youth Health Care, Meppel, the Netherlands. [11]The Feinstein Institute for Medical Research, Cohen Children's Medical Center of NY, Manhasset and New Hyde Park, NY

Address for correspondence: Johnson M. Liu, MD, The Feinstein Institute for Medical Research, Cohen Children's Medical Center of NY, Room 255, New Hyde Park, NY 11040, Jliu3@NSHS.edu

Shwachman-Diamond syndrome (SDS) is an autosomal recessive disorder characterized by pancreatic exocrine insufficiency and bone marrow failure, often associated with neurodevelopmental and skeletal abnormalities. Mutations in the *SBDS* gene have been shown to cause SDS. The purpose of this document is to provide draft guidelines for diagnosis, evaluation of organ and system abnormalities, and treatment of hematologic, pancreatic, dietary, dental, skeletal, and neurodevelopmental complications. New recommendations regarding diagnosis and management are presented, reflecting advances in understanding the genetic basis and clinical manifestations of the disease based on the consensus of experienced clinicians from Canada, Europe, and the United States. Whenever possible, evidence-based conclusions are made, but as with other rare diseases, the data on SDS are often anecdotal. The authors welcome comments from readers.

Introduction

Management: coordinated care model

Shwachman-Diamond syndrome, first described in 1964 (ref [1–3]), is a multi-system disease involving the bone marrow, pancreas, bony skeleton, and other organs. Decisions about patient management are often difficult to make due to the complexity of the clinical phenotype, rarity of the disease and the paucity of large studies. The last report of consensus guidelines for SDS was published in 2002 (ref [4]). With the identification of the *SBDS* gene in 2003 (ref [5]), diagnostic criteria have changed. DNA analysis may lead to the diagnosis of SDS before the full clinical spectrum is present. Informed clinical surveillance and the early findings from experimental models have further highlighted that mutations in SBDS affect a broad spectrum of functions, which has led to a reexamination of the clinical phenotype

and spectrum of the human disease. In particular, neurocognitive manifestations such as learning and behavioral disabilities may be under-recognized. Diversity in how SDS manifests suggests the value of a coordinated multidisciplinary approach to clinical care. Consensus guidelines presented in this document aim to improve health care by highlighting different aspects of SDS and facilitating early diagnosis, prevention and therapy.

General features of SDS

The predominant manifestations of SDS comprise bone marrow failure, pancreatic exocrine dysfunction and skeletal abnormalities.[6–8] In addition, the liver, kidneys, teeth, brain, and immune system may also be affected.[6,9–13] SDS is also associated with a propensity for myelodysplastic syndrome (MDS) and leukemia.[6,9,14–16] SDS is a rare inherited

doi: 10.1111/j.1749-6632.2011.06349.x

marrow failure syndrome with an estimated incidence of 1/76,000 (ref [17]). Although SDS is an autosomal recessive disorder, the ratio of males to females reported in the literature with SDS is 1.7 to 1 (ref [10]).

Hematological manifestations. Neutropenia is the most common hematological abnormality, occurring in nearly all patients. It might be seen in the neonatal period,[6,18] and it can be either persistent or intermittent, fluctuating from severely low to normal levels. In some patients, SDS neutrophils may exhibit defects in migration and chemotaxis.[11,14,19]

Anemia with low reticulocytes occurs in up to 80% of the patients. The red blood cells are usually normochromic and normocytic, but can also be macrocytic.[20] Fetal hemoglobin is elevated in 80% of patients.[21] The anemia is usually asymptomatic. Thrombocytopenia, with platelets less than 150×10^9/l, is variably seen, as are tri-lineage cytopenias. Severe aplasia requiring transfusions has occasionally been reported.[6,22,23]

Bone marrow biopsy usually shows a hypoplastic specimen with increased fat deposition,[6,21] but marrows showing normal or even increased cellularity have also been observed.[10,14] Single-lineage hypoplasia is usually myeloid and occurs in some patients.[9,10] Left-shifted granulopoiesis is a common finding,[6,10] Mild dysplastic changes in the erythroid, myeloid, and megakaryocytic precursors are commonly seen and may fluctuate; however, prominent multilineage dysplasia is less common, and if it occurs, may signify malignant myeloid transformation.

Pancreatic dysfunction, nutrition, and liver disease. Variably severe exocrine pancreatic dysfunction with or without nutrient maldigestion is a hallmark of SDS.[10] Histological specimens of the pancreas have revealed extensive fatty replacement of pancreatic acini with preserved islets of Langerhans and ductal architecture.[3,6] Pancreatic dysfunction is usually diagnosed within the first six months of life and (in 90% of patients) during the first year.[9] Ductular electrolyte and fluid secretion has been shown to remain normal, but the secretion of proteolytic enzymes is severely decreased leading to steatorrhea.[9,24] Spontaneous improvement in pancreatic function can occur in later childhood. By 4 years of age, almost 50% of patients may no longer require pancreatic enzyme supplements as based on evidence of normal fat absorption.[9] Although the causative mechanism is unknown, normalization of fat absorption over the years may remain limited to a subgroup of patients. Despite the relief in subjective symptoms, all patients had a persistent deficit of enzyme secretion in quantitative studies of pancreatic function.[9]

Hepatomegaly is common in young children with SDS. Elevated serum liver enzymes are seen in up to 75% of patients, most often in infants and young children, and tend to resolve with age. Although there are limited longitudinal data, liver disease appears to have little or no long-term clinical consequences.[25] Chronic liver disease has not been observed in a recent series.[26]

Average birth weight is at the 25th percentile. Growth failure with malnutrition is a common feature in the first year of life particularly prior to diagnosis. It is attributable to various factors, including inadequate nutrient intake with or without feeding difficulties, pancreatic insufficiency, and recurrent infections.[6,10] By the first birthday, over half of patients have dropped below the 3rd percentile for both height and weight. After diagnosis, and with appropriate therapy, most children show normal growth velocity, but remain consistently below the 3rd percentile for height and weight.[9]

Other manifestations. SDS-associated bone disease includes skeletal dysplasia[6,10,27–30] and low-turnover osteoporosis.[31] Skeletal dysplasia usually presents with metaphyseal changes in the long bones and costochondral junctions (Fig. 1), but several other less frequent bone anomalies such as supernumerary fingers and syndactyly have also been described.[12,32] In a small cohort, all had some evidence of metaphyseal dysplasia at some point, but the frequency and rate of development are unknown at this time.[27]

Delayed dentition of permanent teeth, dental dysplasia, increased risk of dental caries, and periodontal disease may also occur. On rare occasions, abnormalities of the kidneys, eyes, skin, testes, endocrine pancreas, heart, nervous system, and craniofacial structures have been reported.[6,10,33,34]

How do we diagnose SDS?

Most patients present in infancy with evidence of growth failure, feeding difficulties and/or recurrent

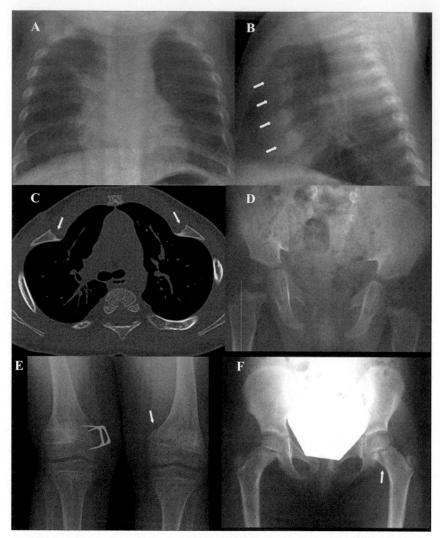

Figure 1. Skeletal radiographic features in SDS. (A and B) Short ribs with marked cupping and widening of the anterior ends (arrows) in a chest X-ray at 11 months. (C) CT slice shows deformed rib cage with short costae and cupping and irregular widening of the costochondral junctions (arrows). (D) Broad pelvis, short iliac notches, valgus position of femoral necks and wide proximal metaphyses of the femora in pelvic X-ray at 11 months. (E) Marked metaphyseal changes with striated bony structure in both hips and the knees at 14 years. Medial hemiepiphyseodesis was performed on the right distal femur due to genum valgum. A stress fracture in the left distal femur (arrow). (F) Broad femoral necks with abnormal metaphyseal structure and a stress fracture in the left femoral neck metaphysis (arrow).

infections. Clinical diagnosis is generally made in the first few years of life but occasionally the diagnosis may be established in older children and even adults. The clinical diagnosis (Table 1) is established by (a) documenting evidence of characteristic exocrine pancreatic dysfunction and hematological abnormalities[10,35,36] and (b) excluding known causes of exocrine pancreatic dysfunction and bone marrow failure.

Attention should be given to ruling out cystic fibrosis (the most common cause of pancreatic insufficiency) with a sweat chloride test, Pearson disease (pancreatic insufficiency and cytopenia, marrow ring sideroblasts and vacuolated erythroid and myeloid precursors), cartilage hair hypoplasia (diarrhea and cytopenia, and metaphyseal chondrodysplasia, and more common in certain isolated populations such as the Amish), and other inherited

Table 1. Clinical and molecular diagnostic criteria

Diagnostic criteria

Clinical diagnosis:

    Fulfill the combined presence of hematological cytopenia of any given lineage (most often neutropenia) and exocrine pancreas dysfunction

Hematologic abnormalities may include:

    a. Neutropenia <1.5 x 109/L on at least 2 occasions over at least 3 months

    b. Hypoproductive cytopenia detected on 2 occasions over at least 3 months

Tests that support the diagnosis but require corroboration:

    a. Persistent elevation of hemoglobin F (on at least 2 occasions over at least 3 months apart)

    b. Persistent red blood cell macrocytosis (on at least 2 occasions over at least 3 months apart), not caused by other etiologies such as hemolysis or a nutritional deficiency

Pancreatic dysfunction may be diagnosed by the following:

    a. Reduced levels of pancreatic enzymes adjusted to age [fecal elastase, serum trypsinogen, serum (iso)amylase, serum lipase]

Tests that support the diagnosis but require corroboration:

    a. Abnormal 72 hr fecal fat analysis

    b. Reduced levels of at least 2 fat-soluble vitamins (A, D, E, K)

    c. Evidence of pancreatic lipomatosis (e.g. ultrasound, CT, MRI, or pathological examination of the pancreas by autopsy)

Additional supportive evidence of SDS may arise from:

    a. Bone abnormalities

    b. Behavioral problems

    c. Presence of a first degree-family member diagnosed before with SDS

Other causes pancreatic insufficiency should be excluded, in particular when the *SBDS* gene mutation analysis is negative

Molecular diagnosis: biallelic *SBDS* gene mutation

    Positive genetic testing for *SBDS* mutations known or predicted to be deleterious, e.g. from protein modeling or expression systems for mutant SBDS

Caveats:

Many situations arise when molecular diagnosis is NOT confirmatory in the presence of clinical symptoms:

No identified mutations (about 10% of cases)

Mutation on one allele only

Gene sequence variations that have unknown or NO phenotypic consequence:

A novel mutation, such as a predicted missense alteration, for which it is not yet possible to predict whether it is disease-causing.

SBDS polymorphisms on one or both alleles. Large population studies may be needed to exclude a sequence polymorphism as a bona fide irrelevant variant.

---

bone marrow failure syndromes (such as dyskeratosis congenita).

Exocrine pancreatic phenotype

The clinical diagnosis of the pancreatic phenotype is challenging as most pancreatic function tests lack sufficient sensitivity and/or specificity. This is complicated by the fact that nearly half of subjects with SDS show improvement in exocrine pancreatic function with advancing age. Exocrine pancreatic reserve loss of 98% must occur before signs and symptoms of maldigestion are present. Thus, 72-hour fecal fat balance studies may be normal despite a significant defect in pancreatic acinar function. The terms pancreatic insufficiency (PI) and pancreatic sufficiency (PS) have been coined to discriminate between subjects with PI, who require pancreatic enzymes supplements with meals and those with PS, who invariably have loss of pancreatic reserve but lack clinical evidence of maldigestion.

For these reasons, alternative approaches are recommended to assess patients with a suspected diagnosis of SDS for evidence of pancreatic dysfunction. Serum pancreatic enzyme concentrations are useful markers of the pancreatic phenotype in patients with SDS.[37] Serum immunoreactive trypsinogen concentrations are low (<6 μg/L) in patients with SDS who have PI. However, in patients with PS, serum trypsinogen concentrations are usually above 6 μg/L, and in one fifth of PS patients, measured concentrations are within the reference range. Thus, a low serum trypsinogen is helpful in identifying the pancreatic phenotype, but a normal value does not exclude impaired exocrine pancreatic function. In contrast, serum pancreatic isoamylase activities in SDS patients are uniformly low at all ages, regardless of pancreatic status or trypsinogen concentration. Unfortunately, serum isoamylase activity cannot be used as a sole marker of the SDS pancreatic phenotype because isoamylase production shows age-dependent postnatal development. Healthy infants have low pancreatic isoamylase concentrations (similar to those observed in SDS), which rise and achieve adult values by approximately three years of age.

To overcome these limitations, serum trypsinogen, isoamylase, and age have been incorporated into a diagnostic rule for the SDS pancreatic phenotype, using the Classification and Regression Tree (CART) analysis of Breiman *et al.*[37] With the exception of patients less than 3 years of age, the diagnostic rule effectively distinguished control individuals from patients with a confirmed clinical diagnosis of SDS.

Several alternative non-invasive approaches to establish or exclude pancreatic dysfunction may be considered, including multi-dimensional imaging (ultrasound, CT, or MRI) for evidence of fatty replacement of the pancreas, and fecal enzyme concentrations of pancreatic elastase or chymotrypsin. Concentrations of fecal elastase less than 200 μg/g stool offer evidence of severe pancreatic dysfunction, and a fecal elastase <100 μg/g is suggestive of maldigestion due to exocrine pancreatic insufficiency. Fecal fat balance studies provide direct evidence of the severity of malabsorption, but as mentioned above, they do not indicate a specifically pancreatic cause if fat malabsorption is found.

The "gold standard" method of directly measuring pancreatic secretion using an intestinal marker perfusion technique to quantify timed collections of pancreatic juice during hormonal stimulation with cholecystokinin and secretin provided useful information concerning the pathophysiology of the exocrine pancreas. However, this complex, invasive test has little role in a clinical setting and is largely used only in research studies. Alternative non-quantitative methods of collecting secretions, including aspiration of pancreatic juice with a duodenoscope or single lumen duodenal tube are not recommended because they show considerable test variability and approximately 25% of PS subjects with low pancreatic reserve may be misclassified as having PI.

Hematologic phenotype

The hematologic phenotype is most frequently characterized by intermittent or persistent neutropenia, but cytopenias of other blood cell lineages are frequently present. Red blood cell macrocytosis, high hemoglobin F, and varying degrees of marrow hypoplasia are also typical findings.

Chromosome breakage studies with diepoxybutane or mitomycin C are recommended to exclude Fanconi anemia, unless the history, physical examination and initial work-up are diagnostic for SDS. Bone marrow aspiration and biopsy are essential for initial evaluation and should include assessment of cellularity, differential, iron stain and cytogenetics. Bone marrow cytogenetic finding of i(7q) or del(20q) is highly associated with SDS. Virology studies (e.g. Epstein–Barr virus, cytomegalovirus, and B19 parvovirus) may be pursued as clinically indicated to exclude other causes of bone marrow suppression and a failure to thrive.

Skeletal phenotype

When present in association with hematologic or pancreatic abnormalities, characteristic skeletal abnormalities are strongly suggestive of SDS. SDS bone dysplasia is characterized by short stature, delayed appearance but subsequent normal development of secondary ossification centers, and by variable metaphyseal widening and irregularity that is most often seen in the ribs in early childhood and in the proximal and distal femora later in childhood and adolescence.[10,27] Rarely, skeletal involvement may be extremely severe with generalized bone abnormalities.[38] Although metaphyseal changes often become undetectable and clinically insignificant over time, they may also progress and result in limb deformities, most commonly at the hips and the knees, or stress fractures of the femoral necks (Fig. 1).[27] In addition to metaphyseal chondrodysplasia, SDS associates with early-onset low-turnover osteoporosis characterized by low bone mass and vertebral fragility fractures.[31]

Other clinical findings
Short stature with or without malnutrition is also a common feature of SDS. Hepatomegaly with mild to moderate biochemical abnormalities of the liver are common findings in infants and young children with SDS.

Molecular testing
As the clinical diagnosis of SDS is usually difficult and patients may present at a stage when no clinical pancreatic insufficiency is evident, it is advisable to test most or all suspected cases for mutations in the *SBDS* gene (Table 1). It is noteworthy that about 10% of the SDS patients may be negative for mutations, and that *de novo SBDS* mutations have been identified in some families.

How to monitor a patient after a diagnosis is made?

Recommended baseline testing are listed in Table 2.

Hematology
Hematological evaluation should include complete blood count (CBC), mean corpuscular volume, peripheral blood smear, differential, reticulocyte count, fetal hemoglobin level and coagulation tests in case of clinical bleeding symptoms. If the diagnosis of SDS is suspected or confirmed, bone marrow aspirate smear, biopsy, and cytogenetic evaluation is recommended as a baseline examination (see section IV for further discussion).

Complete blood count is a basic parameter that needs to be monitored: CBCs should be considered every 3–6 months in stable patients. Any clinical complications, including recurrent infections, bruising, asthenia or pallor may require a CBC between scheduled examinations. The purpose of the routine CBC is to determine the baseline profile of the patients, to assess the risk for infections and possibly to detect particular features related to nutritional deficits, such as iron or folate deficiency and to detect evolving marrow abnormalities such

as severe marrow failure, myelodysplastic syndrome (MDS), or leukemia.

When infections regularly recur, immunoglobulin levels and post-vaccination antibodies should be screened to exclude an associated immunodeficiency.

Systematic evaluation of neutrophil chemotaxis is not considered a necessity in the usual follow up of patients.

Pancreas
Once the diagnosis of SDS is suspected or established, objective testing for assessment of pancreatic function status is recommended. To determine PS or PI status, serum trypsinogen concentration offers useful screening information:

(a) If values are undetectable or low, a 72-hour fat balance study may be done to confirm PI status. Since most newly diagnosed subjects are infants or children, careful documentation of ingested fat (and other macronutrients) will enable determination of coefficient of fat absorption as well as provide insight into total calorie intake.

(b) If values are 6 μg/L or above, PS status should be confirmed by 72-hour fat balance study as described. Recent studies in patients with cystic fibrosis have, however, shown that duplicate measurements of the coefficient of fat absorption often show wide variation.

(c) Measurement of fecal elastase or chymotrypsin is widely used in Europe as an alternative indicator of pancreatic insufficiency, although it has not been validated in a large series of SDS patients. It has the theoretical advantage of being a specific test of pancreatic function, whereas fat absorption can of course be abnormal in non-pancreatic disorders such as celiac disease.

Baseline fat soluble vitamin levels (A, D, E) and prothrombin time, as a surrogate marker for vitamin K status, should be done. Low values should be correlated with results of pancreatic function testing and in patients with PI, should be repeated approximately one month after instituting enzyme replacement therapy. Persistently low levels in the face of good compliance with enzyme therapy will require fat-soluble vitamin supplements. Fat-soluble vitamins should be monitored on at least a yearly basis, and may include (vitamin K-dependent) coagulation parameters when clinical symptoms are present.

Table 2. Clinical tests at diagnosis and at follow-up

	At Diagnosis	At Follow-up
Genetics		
SBDS gene mutation (test may be offered to family member hematopoietic stem cell transplant donors)	Yes	Yes, if not done at diagnosis
Genetic counselling (molecular test may be offered to family members for screening of carriers)		
Hematology and immunology		
CBC	Yes	2–4 times / year
Bone marrow aspirate and biopsy	Yes	Every 1 to 3 years or as clinically indicated
Fe, folate, B12 levels	Yes	
Hb F levels	Yes	As clinically indicated
IgG, IgA, IgM levels	Yes	–
Post vaccination serology	–	As clinically indicated
Lymphocyte phenotype	–	As clinically indicated
HLA testing	As clinically indicated	As clinically indicated
Gastroenterology		
Pancreatic enzymes (choice based on local availability: serum trypsinogen, isoamylase, 72-hour fat balance test, elastase, etc.)	Yes	
Fat-soluble vitamins A, D, E, and prothrombin time (surrogate for vitamin K)	Yes	1 mo after pancreatic enzyme therapy, then 1-2 times/ year
Other vitamins and micronutrients	–	As clinically indicated
Liver biochemistry panel	Yes	As clinically indicated
Pancreatic imaging	Ultrasound (abdomen)	
Endoscopy	As clinically indicated	
Skeletal system, growth		
Growth evaluation: height, weight and head circumference	Yes	Yearly at follow-up
Skeletal survey	Yes	As clinically indicated
Densitometry		Baseline study: once during prepuberty Follow-up study: once during puberty, then as clinically indicated
Oral and dental care	Yes	Once per year and when clinically indicated

Continued

Table 2. *Continued*

	At Diagnosis	At Follow-up
Development and neuropsychological evaluation	Yes	Standardized developmental screening measure: Infancy/-preschool age Neuropsychological assessment of domains: At ages 6–8, 11–13, 15–17
		Intellectual abilities
		Attention including working memory, sustained attention and divided/dual attention
		Higher order language
		Visual-motor integration and speed
		Executive functioning
		Academic achievement
		Behaviour (self report and parent proxy)
		Adaptive Functioning (parent proxy)

There are no published guidelines on dosing of pancreatic enzyme supplements in SDS patients with PI. Furthermore, there are few published data demonstrating efficacy of enzyme replacement therapy. For this reason, published treatment guidelines for subjects with cystic fibrosis may be considered.[39]

Nutritional status

Newly diagnosed infants with SDS are commonly malnourished. Therefore, careful baseline assessment of height and weight and anthropometric measures are recommended. Once appropriate therapy is introduced, malnutrition should be corrected by one year of age.

Bone

Skeletal survey is recommended at the time of the diagnosis. The follow-up is based on individual clinical and radiological findings. For biochemical assessment and bone mineral density evaluation, see section on bone abnormalities.

Dental

Annual reviews—ideally by a dentist experienced in orthodontic approaches and/or periodontal disease—are generally recommended.

Neurodevelopment

A characteristic pattern of learning and behavioral difficulties is common in SDS.[40] It is therefore important to monitor and support neurodevelopment. Standardized developmental checklists should be used routinely to assess infant, toddler and preschooler development with referrals to special-ists (e.g., speech and language therapy, occupational therapy, developmental pediatrician, developmental psychologist) as needed. Serial neuropsychological assessments are indicated, at minimum, when a child is approximately 6, 12, and 15 years of age to correspond with brain development and changes in expectations at school.

Hematological complications

Definition of hematological complications

While neutropenia (even severe) is a typical feature of SDS, anemia (<7 g/dl or 4.3 mmol/L or if symptomatic) and thrombocytopenia ($<20 \times 10^9$/L or if symptomatic) are additional complications that require prompt evaluation and medical decision.

Classification of the different forms of marrow failure in SDS is complex and poorly understood. In general, cytogenetic studies should be performed concurrently with morphology studies. Aplastic anemia (hypoproliferative cytopenia without dysplastic morphology and usually without clonal evolution) and myelodysplastic syndrome (cytopenia with dysplastic morphology and clonal evolution) represent the two main categories of complications. However, most of the common scenarios seen in SDS differ from the standard definitions established by World Health Organization (WHO) criteria,[41] because the bone marrow morphology from SDS patients often bears mild dysplastic changes in the erythroid, myeloid and megakaryocytic series, even in the absence of clonal cytogenetic abnormalities.

Aplastic anemia

Aplastic anemia can be divided into moderate and severe subcategories.[42,43] Severe disease is defined by depression in two of three blood counts (reticulocytes <40,000/μL, platelets <20,000/μL, neutrophils <500/μL) in the presence of a hypocellular bone marrow biopsy (<25% cellularity or <50% cellularity and <30% hematopoietic cells) without significant fibrosis. Moderate disease is defined as failure to meet the criteria for severe disease but with at least two diminished blood counts (reticulocytes <40,000/μL, platelets <40,000/μL, neutrophils <1,500/μL) with a hypocellular bone marrow biopsy.

The diagnosis of aplastic anemia is usually, but not always, considered in the absence of clonal marrow cytogenetic abnormalities (CMCA). Aplastic anemia may be transient (lasting less than 3 months) or may persist past 3 months, becoming clinically significant (J. Donadieu, unpublished data).

Clonal marrow cytogenetic abnormality

Clonal marrow cytogenetic abnormality (CMCA) is defined by: two or more bone marrow cells (out of twenty) with gain of the same chromosome or cytogenetic abnormality or three or more cells with loss of the same chromosome, as detected by G-banding; or a cytogenetic abnormality detected by fluorescence *in situ* hybridization (FISH) analysis in higher frequency than the reference values of the lab, as well as higher than in the concurrently tested control sample.

Diagnostic criteria for MDS and AML

The critical component for MDS is dysplastic morphology, as defined by the WHO.[41] Published criteria for MDS in children include two out of the following three items: chronic trilineage cytopenia, prominent bi-lineage cytopenia, clonal marrow cytogenetic abnormality, marrow myeloblast count between 5–29%.[44,45] However, since cytogenetic abnormalities as well as mild dysplastic features occur in some SDS patients without progression to AML, the markers that discriminate MDS from the aplastic phase are still debatable. AML is defined by a marrow myeloblast count of ≥20% (WHO)[41] or ≥30% (French American British classification).[46]

There are two current classification systems for pediatric MDS,[44,45] but the prognostic significance of the systems has not yet been studied. A literature review[47] reveals that subjects with SDS commonly show clonal marrow cytogenetic abnormalities (CMCA), MDS or AML. Among those identified with CMCA/MDS in childhood, approximately 50% progressed to overt leukemia over a range of 1 to 37 years. Remarkably, males constituted 68% and 92% of all subjects with CMCA/MDS and leukemia, respectively.

The bone marrow cytogenetic abnormalities i(7q) and del(20q) are quite common in SDS, occur less frequently in other malignancies or marrow failure syndromes, and can regress spontaneously.[16,48] These specific cytogenetic changes may be relatively specific for SDS and, in isolation, may not be an absolute harbinger of malignancy. In general, cytogenetic abnormalities of unclear clinical significance should be interpreted in the context of the marrow morphology and blast count.[14,15,48–54] Of these patients, some developed severe aplasia, while others progressed to more severe MDS/AML. SDS patients may also present with MDS at the stage of refractory cytopenia with dysplasia[14,15,48–54] or with excess blasts, some of whom progress to AML.

Various types of AML have been described in SDS patients: AML-M0, M2, M4, M5, and M6. Acute lymphoblastic leukemia and juvenile myelomonocytic leukemia were rare. AML-M6 was particularly common in SDS, occurring in about 30% of cases with classifiable leukemia. Malignant myeloid transformation into MDS and AML in SDS patients while on G-CSF therapy has been reported,[49,55,56] but the causal relationship is unproven. SDS-related leukemia carries a poor prognosis if treated with chemotherapy alone. However, due to the improving outcome of stem cell transplantation in patients over the past years, the prognosis of SDS with secondary leukemia has improved accordingly, but data are still limited.

Surveillance

In cases presenting with severe pancytopenia, bone marrow aspirate, biopsy, and cytogenetic examination are mandatory. However, the indications for routine bone marrow smear and bone marrow cytogenetics are controversial. To date, in the absence of severe cytopenia, bone marrow cytogenetic analysis has not generally been predictive of outcome. However, non-i(7q) abnormalities of chromosome 7, particularly monosomy 7, are associated with poor outcomes and may present with advanced

MDS/AML or progress from earlier stages of MDS. In addition, systematic bone marrow cytogenetic examination may have a role in surveillance in patients receiving long-term therapy with granulocyte colony-stimulating factor (G-CSF, see below).

In summary, bone marrow aspirate and biopsy are recommended at the time of diagnosis of SDS, in cases of CBC changes, and annually in patients who are treated with G-CSF therapy. In a patient with stable clinical status and complete blood counts (not on G-CSF), a bone marrow aspirate with cytogenetic examination can be proposed routinely every 1–3 years.

Treatment of hematologic and infectious complications

Cytopenias
Thrombocytopenia and anemia may require respective chronic transfusions, with institution of an iron-chelation program as clinically indicated. If transfusions are indicated, blood products need to be irradiated.

Granulocyte colony stimulating factor
The majority of patients do not need granulocyte colony stimulating factor (G-CSF) due to the low incidence of infections. Chronic use of G-CSF should be considered for recurrent invasive bacterial and/or fungal infections in the presence of severe neutropenia. G-CSF given for profound and persistent neutropenia has been effective in inducing a clinically beneficial neutrophil response. Patients may respond to an intermittent schedule with low doses of G-CSF (e.g. 2–3 μg/kg every 3 days) or may require higher doses continuously. The aim of long-term G-CSF treatment is not to obtain normal hematological parameters but to prevent infections. In cases of G-CSF resistance, associated with severe infections, hematopoietic stem cell transplantation (HSCT) should be considered.

Androgens
Data are scarce regarding response rates to androgens in SDS patients. A few patients have received androgens, and responses have been reported. However, androgens are generally not recommended as first line therapy for severe bone marrow failure in SDS. Underlying liver abnormalities seen in SDS may lead to higher liver toxicity than that seen in Fanconi anemia. The use of androgens should probably be reserved for patients who do not have severe

bone marrow failure, and for whom an HSCT donor is unavailable.

Prevention and treatment of infections
Patients with acute infectious episodes, suggested by fever or any acute symptoms need to be evaluated urgently. Some patients can be treated with oral antibiotics, while patients with severe neutropenia or those suspected to have severe infections should be hospitalized and treated with intravenous antibiotics with broad-spectrum coverage until improvement. G-CSF treatment should also be considered during infections in patients with severe neutropenia. In cases of recurrent infections or severe chronic stomatitis with profound neutropenia, long-term G-CSF therapy may be considered (see above).

Bleeding episodes
In the presence of thrombocytopenia or low vitamin K-dependent coagulation factors, bleeding may occur. Mild to moderate bleeding episodes can be treated with local measures (xylometazoline 0.05% nose spray), tranexamic acid, or aminocaproic acid. When coagulation is affected by low vitamin K and/or, rarely, abnormal liver function, vitamin K should be administered. Platelet transfusions are indicated in an SDS patient with severe bleeding and thrombocytopenia. Prophylactic administration of platelets should be considered for patients with platelet counts of $<10 \times 10^9/L$ or for those with a known tendency to have significant bleeding episodes.

For surgery or invasive procedures, platelets should be transfused as clinically indicated. When known or suspected coagulation defects are present, infusion of fresh frozen plasma or plasma-derived coagulation products (such as prothrombin complex, containing factors II, VII, IX, and X) may be indicated.

Female patients suffering from blood loss during menstruation may benefit from pharmacologic treatment to induce amenorrhea.

MDS and AML: chemotherapy
In MDS secondary to SDS, standard chemotherapy regimens are not indicated and an attempt should be made to provide HSCT on an urgent basis. High dose chemotherapy is therefore mainly indicated for conditioning prior to HSCT.

Standard chemotherapy for AML can be effective to temporarily control the disease. However, chemotherapy alone has been unsuccessful in obtaining a prolonged complete remission in SDS. Therefore, due to a high risk of persistent aplasia, an urgent search for a related or unrelated donor for HSCT should be initiated and minimal chemotherapy to provide interim disease control should be considered.

Hematopoietic stem cell transplantation

Indications for HSCT. The criteria for considering patients for HSCT (related or alternative) include:

(a) Severe cytopenia [hemoglobin <7 g/L (4.3mmol/L), absolute neutrophil count $< 0.5 \times 10^9$/L with recurrent infections, platelet count $<20 \times 10^9$/L]
(b) MDS with excess blasts
(c) Overt leukemia

In cases of frank leukemia, the patient may be started on chemotherapy to reduce tumor load before HSCT, but an effort to find a donor should be made at the time of diagnosis because of the high risk of therapy-related aplasia.

In considering the indications for HSCT, one should also allow for the possibility of spontaneous recovery from aplasia. Depending on the level of potential immediate risks of the severe cytopenia, a monitoring period of up to 3 months can be considered while concurrently initiating the process of HLA typing and donor search.

Conditioning regimen and GVHD prophylaxis. At present, HSCT provides the only curative option for the hematological complications in SDS. Reported cases of SDS patients who have undergone HSCT include no more than 80 patients worldwide.[57,58] Many different conditioning/supportive regimens in small groups of patients render general conclusions and recommendations difficult. Globally, it appears that the results depend on the type of donor (genotypically identical donor transplants better than matched unrelated donor or MUD transplants) in almost all reports. However, the indications for HSCT also appear to be a clear determinant of survival. The survival of patients receiving a transplant for aplastic anemia is about 80%, while the survival of patients receiving a transplant for MDS or acute leukemia remains between 30 and 40%. This disparity is likely due in part to differences in the ages of recipients, because aplastic anemia is usually a complication in the first decade of life, whereas MDS/AML is more likely a complication of

the second or third decade (younger patients generally have better outcomes following HSCT). Most data have been collected over the past 20 years, and current results may be more promising due to better standards for donor searches and treatment of complications.

Complications from chemotherapy or HSCT are more common in SDS patients than in patients with idiopathic blood dyscrasias. In a review of 36 patients with SDS who had been treated with chemotherapy alone[9,14,20,33,49–51,59–63] or with HSCT with or without irradiation, 83% died from complications related to the therapy, including prolonged severe aplasia, infections, cardiotoxicity, neurological and renal complications, veno-occlusive disease, pulmonary disease, post-transplant graft failure, and GVHD. Toxicity, particularly cardiac toxicity,[64] seems more frequent if the indication is MDS/acute leukemia rather than aplastic anemia. Recently, an attenuated conditioning regimen has been proposed in order to limit toxicity.[65,66]

Treatment of pancreatic dysfunction, nutrition and liver disease

Pancreatic enzymes

The clinical response to enzyme treatment in patients with SDS, in contrast to patients with cystic fibrosis for whom there may be additional intestinal factors, is usually excellent, although growth may continue to be restricted for skeletal reasons. The natural history of SDS suggests that pancreatic function may improve to sufficient levels in many patients to allow them to discontinue enzyme supplementation as they become older. The pancreatic status of all patients should therefore be reassessed from time to time, according to their clinical progress.

Once the diagnosis is made, and steatorrhea confirmed, pancreatic enzyme replacement should be started. The initial dose should be 2,000 lipase units/Kg body weight/day. The dosing guidelines for subjects with cystic fibrosis disease (maximum 10,000 lipase units/kg body weight/day) should be followed.[39] Pancreatin is taken with all meals and snacks that contain protein, fat or complex carbohydrates. In children with persistent fat malabsorption despite optimal dose of replacement, an H_2-receptor antagonist or proton pump inhibitor may be given in addition. Higher requirements of pancreatic enzymes should alert the clinician to the possibility of a concomitant unrelated enteropathy.

Enteric-coated enzyme preparations prevent gastric acid-peptic degradation and therefore deliver a higher concentration of enzymes to the intestine than uncoated preparations. The capsules should be swallowed whole, without chewing. If the patient cannot swallow capsules, they can be opened and the enteric-coated granules mixed with milk, juice or pureed fruit. The resulting mixture should be swallowed immediately without chewing. Pancreatin is inactivated at high temperatures, and excessive heat should be avoided when the granules are mixed with liquids or food.

Vitamin supplements

Blood levels of fat-soluble vitamins should be measured every 6 to 12 months in young children, and supplementary therapy started if values are low. It is important to ensure compliance with pancreatic enzyme supplementation, as deficiencies of these vitamins are an indirect marker of fat malabsorption

Dietary advice and surveillance

Height and weight should be documented at every clinic visit. All patients should receive an evaluation by a dietitian. Poor appetite and behavioral feeding difficulties are common. Such children should have a careful psychology assessment and support offered to the family by a clinical psychologist.

If oral intake is suboptimal nutritional supplements should be considered. If there are ongoing concerns about poor weight gain despite adequate pancreatic enzyme replacement therapy, it may be necessary to assess the child for other causes or conditions such as gastro-esophageal reflux, food allergy and enteropathy.[67]

In severe cases of persistent failure to thrive or feeding difficulties, as a last resort a gastrostomy insertion can be considered to allow overnight feeding, but weaning should be attempted once the patient is stable.

Treatment of dental complications

Oral and dental problems are common in children with SDS[68]. Ulceration of the oral mucosa can be associated with neutropenia. The frequency and severity of the ulceration is variable. Enamel defects have been noted, in both the deciduous and permanent dentitions. Areas of faulty mineralization of the dental surface can lead to decay and can be severe in some cases. Gastric acid reflux can lead to tooth surface loss or erosion. Regular dental care and appropriate advice from an early age are crucial to minimize these oral and dental problems.

Treatment of bone abnormalities

Treatment and follow-up

Bone deformities due to metaphyseal chondrodysplasia, usually located at the hips or the knees, may require orthopedic consultation and surgical interventions. Low-turnover osteoporosis may result from a primary defect in bone metabolism that is related to the bone marrow dysfunction and neutropenia. Efforts should be made to optimize general preventive measures such as nutrition and intake of fat-soluble vitamins, as well as to promote weight-bearing exercise. Supplementation with vitamin D (in addition to other fat-soluble vitamins) and calcium should be commenced if dietary intakes are not sufficient. It is presently unknown whether bisphosphonates, anti-resorptive agents used to treat postmenopausal high-turnover osteoporosis, are safe and efficacious in SDS osteoporosis. Optimal treatment for SDS osteoporosis remains to be established.

Radiography and bone densitometry. Assessment of bone dysplasia (Tables 2 and 3): at diagnosis, radiographic skeletal survey; follow-up based on individual clinical and radiographic findings, X-rays for detection of deformities or stress fractures (hips, knees). Assessment of osteoporosis: bone densitometry by DXA, at prepuberty (baseline study), during pubertal years, postpubertal follow-up studies based on individual findings (low BMD, vertebral compressions, multiple peripheral fractures). Caution should be exercised when interpreting DXA results in patients with SDS; small body size and delayed pubertal development affect BMD results.

Biochemistry. Serum 25-OH-vitamin D and plasma parathyroid hormone (PTH) should be monitored as part of routine follow-up and maintained within normal limits after the diagnosis.

Neurodevelopmental consequences and support

Deficits in cognitive abilities across numerous domains of functioning are evident in the majority of individuals with SDS at varying levels of severity

Table 3. Longitudinal changes in skeletal phenotype in SDS

When?	What?	Where?
Infancy and early childhood	Delayed appearance of secondary ossification centers	Wrist, hand, femur
	Wide, irregular metaphyses	Ribs, wrist
	Osteopenia, Wormian bones	Tubular bones, skull
Mid-childhood	Slow development of secondary ossification centers	Wrist, hand, femur
	Irregularity and sclerosis of metaphyses	Femur
	Osteopenia	Tubular bones, spine
Late childhood/ puberty	Irregularity, sclerosis and asymmetrical growth of metaphyses	Femur
	Stress fractures, deformity	Femur
	Compression fractures	Spine
Adulthood	Compression fractures	Spine

indicating heterogeneity. Parental report indicates that over 50% of children experience delayed language development.[6,40] Below average intellectual reasoning abilities are also evident[6,40,69,70] with approximately 1 in 5 meeting the diagnostic criteria for an intellectual disability (i.e., IQ < 2nd percentile).[40] Difficulties in visual reasoning and visual-motor integration,[40,70] higher order language functioning (e.g. understanding figurative expressions, knowledge of synonyms), executive problem solving and attention have also been documented.[40]

Significant behavioral issues are commonly reported. In a study of 32 children / adolescents (ages 6 through 17),[40] 19 percent had prior diagnosis of attention deficit hyperactivity disorder, pervasive developmental disorder or oppositional defiant disorder while an additional 31 percent were reported to have some combination of inattention, restless, impulsivity, and oppositional behavior. In addition, on behavioral rating scales, parents indicated a heightened frequency of attention problems (50%) and social problems (34%). The neurocognitive deficits have been found to be independent of pancreatic involvement, otitis media, having a chronic illness, family environment, and age.[40] Given the structural abnormalities that are evident on neuro-imaging of the brain,[71-73] neurocognitive and neurobehavioral issues are likely the consequences of SBDS gene dysfunction on the brain.

Assessment, monitoring, and treatment

In order to maximize ongoing development, comprehensive assessments using standardized tests and clinical observation to monitor cognitive, behav-ioral, social, and adaptive functioning are warranted from time of diagnosis through to adulthood. Specifically, during the infancy/pre-school period (diagnosis to 4 years of age), it is advised that comprehensive developmental checklists be used so that referrals to specialists (i.e., speech and language therapist, occupational therapist, developmental pediatrician, developmental psychologist), assessment and intervention can occur at the earliest sign of possible issues. In addition, it is recommended that serial neuropsychological assessments be completed to coincide with key stages of brain maturation, namely 6–8, 11–13, and 15–17 years of age. These age groups also parallel changes in expectations in learning at school. Assessments should include evaluation of intellectual abilities, attention (working memory, sustained attention, and divided/dual attention), higher order language, visual perception, visual-motor functioning, executive skills, academic readiness/achievement, behavior, and functional independence. The identification of an individual's strengths and weaknesses, consequently leads to individualize recommendations for intervention, which are reviewed and adapted at the follow-up assessment at the next critical stage of development. Counselling for parents should parallel the neuropsychological assessments of their child to support them in enhancing interactions with, and in developing realistic expectations for, their child.

Conflicts of interest

The authors declare no conflicts of interest.

References

1. Shwachman, H., L.K. Diamond, F.A. Oski & K.T. Khaw. 1964. The syndrome of pancreatic insufficiency and bone marrow dysfunction. *J. Pediatr.* **65:** 645–663.

2. Nezelof, C. & M. Watchi. 1961. [Lipomatous congenital hypoplasia of the exocrine pancreas in children. (2 cases and review of the literature)]. *Arch. Fr. Pediatr.* **18:** 1135–1172.

3. Bodian, M., W. Sheldon & R. Lightwood. 1964. Congenital Hypoplasia of the exocrine pancreas. *Acta. Paediatr.* **53:** 282–293.

4. Rothbaum, R., J. Perrault, A. Vlachos, *et al.* 2002. Shwachman-Diamond syndrome: report from an international conference. *J. Pediatr.* **141:** 266–270.

5. Boocock, G.R., J.A. Morrison, M. Popovic, *et al.* 2003. Mutations in SBDS are associated with Shwachman-Diamond syndrome. *Nat. Genet.* **33:** 97–101.

6. Aggett, P.J., N.P. Cavanagh, D.J. Matthew, *et al.* 1980. Shwachman's syndrome. A review of 21 cases. *Arch. Dis. Child.* **55:** 331–347.

7. Burke, V., J.H. Colebatch, C.M. Anderson & M.J. Simons. 1967. Association of pancreatic insufficiency and chronic neutropenia in childhood. *Arch. Dis. Child.* **42:** 147–157.

8. Pringle, E.M., W.F. Young & E.M. Haworth. 1968. Syndrome of pancreatic insufficiency, blood dyscrasia and metaphyseal dysplasia. *Proc. R. Soc. Med.* **61:** 776–778.

9. Mack, D.R., G.G. Forstner, M. Wilschanski, *et al.* 1996. Shwachman syndrome: exocrine pancreatic dysfunction and variable phenotypic expression. *Gastroenterology* **111:** 1593–1602.

10. Ginzberg, H., J. Shin, L. Ellis, *et al.* 1999. Shwachman syndrome: phenotypic manifestations of sibling sets and isolated cases in a large patient cohort are similar. *J. Pediatr.* **135:** 81–88.

11. Dror, Y., H. Ginzberg, I. Dalal, *et al.* 2001. Immune function in patients with Shwachman-Diamond syndrome. *Br. J. Haematol.* **114:** 712–717.

12. Dror, Y., P. Durie, P. Marcon & M.H. Freedman. 1998. Duplication of distal thumb phalanx in Shwachman-Diamond syndrome. *Am. J. Med. Genet.* **78:** 67–69.

13. Dror, Y. & M.H. Freedman. 2002. Shwachman-diamond syndrome. *Br. J. Haematol.* **118:** 701–713.

14. Smith, O.P., I.M. Hann, J.M. Chessells, *et al.* 1996. Haematological abnormalities in Shwachman-Diamond syndrome. *Br. J. Haematol.* **94:** 279–284.

15. Dror, Y., J. Squire, P. Durie & M.H. Freedman. 1998. Malignant myeloid transformation with isochromosome 7q in Shwachman-Diamond syndrome. *Leukemia* **12:** 1591–1595.

16. Dror, Y., P. Durie, H. Ginzberg, *et al.* 2002. Clonal evolution in marrows of patients with Shwachman-Diamond syndrome: a prospective 5-year follow-up study. *Exp. Hematol.* **30:** 659–669.

17. Goobie, S., M. Popovic, J. Morrison, *et al.* 2001. Shwachman-Diamond syndrome with exocrine pancreatic dysfunction and bone marrow failure maps to the centromeric region of chromosome 7. *Am. J. Hum. Genet.* **68:** 1048–1054.

18. Kuijpers, T.W., E. Nannenberg, M. Alders, *et al.* 2004. Congenital aplastic anemia caused by mutations in the SBDS gene: a rare presentation of Shwachman-Diamond syndrome. *Pediatrics* **114:** e387–e391.

19. Rothbaum, R.J., D.A. Williams & C.C. Daugherty. 1982. Unusual surface distribution of concanavalin A reflects a cytoskeletal defect in neutrophils in Shwachman's syndrome. *Lancet* **2:** 800–801.

20. Woods, W.G., W. Krivit, B.H. Lubin & N.K. Ramsay. 1981. Aplastic anemia associated with the Shwachman syndrome. In vivo and in vitro observations. *Am. J. Pediatr. Hematol. Oncol.* **3:** 347–351.

21. Dror, Y. & M.H. Freedman. 1999. Shwachman-Diamond syndrome: An inherited preleukemic bone marrow failure disorder with aberrant hematopoietic progenitors and faulty marrow microenvironment. *Blood* **94:** 3048–3054.

22. Tsai, P.H., I. Sahdev, A. Herry & J.M. Lipton. 1990. Fatal cyclophosphamide-induced congestive heart failure in a 10-year-old boy with Shwachman-Diamond syndrome and severe bone marrow failure treated with allogeneic bone marrow transplantation. *Am. J. Pediatr. Hematol. Oncol.* **12:** 472–476.

23. Barrios, N., D. Kirkpatrick, O. Regueira, *et al.* 1991. Bone marrow transplant in Shwachman Diamond syndrome. *Br. J. Haematol.* **79:** 337–338.

24. Hill, R.E., P.R. Durie, K.J. Gaskin, *et al.* 1982. Steatorrhea and pancreatic insufficiency in Shwachman syndrome. *Gastroenterology* **83:** 22–27.

25. Cipolli, M. 2001. Shwachman-Diamond syndrome: clinical phenotypes. *Pancreatology* **1:** 543–548.

26. Toiviainen-Salo, S., P.R. Durie, K. Numminen, *et al.* 2009. The natural history of Shwachman-Diamond syndrome-associated liver disease from childhood to adulthood. *J. Pediatr.* **155:** 807–811.

27. Makitie, O., L. Ellis, P.R. Durie, *et al.* 2004. Skeletal phenotype in patients with Shwachman-Diamond syndrome and mutations in SBDS. *Clin. Genet.* **65:** 101–112.

28. Taybi, H., A.D. Mitchell & G.D. Friedman. 1969. Metaphyseal dysostosis and the associated syndrome of pancreatic insufficiency and blood disorders. *Radiology* **93:** 563–571.

29. Danks, D.M., R. Haslam, V. Mayne, *et al.* 1976. Metaphyseal chondrodysplasia, neutropenia, and pancreatic insufficiency presenting with respiratory distress in the neonatal period. *Arch. Dis. Child.* **51:** 697–702.

30. Labrune, M., J.P. Dommergues, C. Chaboche & J.J. Benichou. 1984. [Shwachman's syndrome with neonatal thoracic manifestations]. *Arch. Fr. Pediatr.* **41:** 561–563.

31. Toiviainen-Salo, S., M.K. Mayranpaa, P.R. Durie, *et al.* 2007. Shwachman-Diamond syndrome is associated with low-turnover osteoporosis. *Bone* **41:** 965–972.

32. Dror, Y., H. Ginzberg, I. Dalal, *et al.* 2001. Immune function in patients with Shwachman-Diamond syndrome. *Br. J. Haematol.* **114:** 712–717.

33. Dokal, I., S. Rule, F. Chen, *et al.* 1997. Adult onset of acute myeloid leukaemia (M6) in patients with Shwachman-Diamond syndrome. *Br. J. Haematol.* **99:** 171–173.

34. Savilahti, E. & J. Rapola. 1984. Frequent myocardial lesions in Shwachman's syndrome. Eight fatal cases among 16 Finnish patients. *Acta. Paediatr. Scand.* **73:** 642–651.

35. Thornley, I., Y. Dror, L. Sung, *et al.* 2002. Abnormal telomere shortening in leucocytes of children with Shwachman-Diamond syndrome. *Br. J. Haematol.* **117:** 189–192.

36. Woloszynek, J.R., R.J. Rothbaum, A.S. Rawls, *et al.* 2004. Mutations of the SBDS gene are present in most patients

with Shwachman-Diamond syndrome. *Blood* **104:** 3588–3590.

37. Ip, W.F., A. Dupuis, L. Ellis, *et al.* 2002. Serum pancreatic enzymes define the pancreatic phenotype in patients with Shwachman-Diamond syndrome. *J. Pediatr.* **141:** 259–265.

38. Nishimura, G., E. Nakashima, Y. Hirose, *et al.* 2007. The Shwachman-Bodian-Diamond syndrome gene mutations cause a neonatal form of spondylometaphysial dysplasia (SMD) resembling SMD Sedaghatian type. *J. Med. Genet.* **44:** e73.

39. Borowitz, D., K.A. Robinson, M. Rosenfeld, *et al.* 2009. Cystic Fibrosis Foundation evidence-based guidelines for management of infants with cystic fibrosis. *J. Pediatr.* **155:** S73–S93.

40. Kerr, E.N., L. Ellis, A. Dupuis, *et al.* 2010. The behavioral phenotype of school-age children with shwachman diamond syndrome indicates neurocognitive dysfunction with loss of Shwachman-Bodian-Diamond syndrome gene function. *J. Pediatr.* **156:** 433–438.

41. Vardiman, J.W., J. Thiele, D.A. Arber, *et al.* 2009. The 2008 revision of the World Health Organization (WHO) classification of myeloid neoplasms and acute leukemia: rationale and important changes. *Blood* **114:** 937–951.

42. Camitta, B.M. 1988. Criteria for severe aplastic anaemia. *Lancet* **1:** 303–304.

43. Young, N., P. Griffith, E. Brittain, *et al.* 1988. A multicenter trial of antithymocyte globulin in aplastic anemia and related diseases. *Blood* **72:** 1861–1869.

44. Mandel, K., Y. Dror, A. Poon & M.H. Freedman. 2002. A practical, comprehensive classification for pediatric myelodysplastic syndromes: the CCC system. *J. Pediatr. Hematol. Oncol.* 24: 596–605.

45. Hasle, H., C.M. Niemeyer, J.M. Chessells, *et al.* 2003. A pediatric approach to the WHO classification of myelodysplastic and myeloproliferative diseases. *Leukemia* **17:** 277–282.

46. Bennett, J.M., D. Catovsky, M.T. Daniel, *et al.* 1982. Proposals for the classification of the myelodysplastic syndromes. *Br. J. Haematol.* **51:** 189–199.

47. Dror, Y. 2005. Shwachman-Diamond syndrome. *Pediatr. Blood Cancer* **45:** 892–901.

48. Smith, A., P.J. Shaw, B. Webster, *et al.* 2002. Intermittent 20q- and consistent i(7q) in a patient with Shwachman-Diamond syndrome. *Pediatr. Hematol. Oncol.* **19:** 525–528.

49. Davies, S.M., J.E. Wagner, T. DeFor, *et al.* 1997. Unrelated donor bone marrow transplantation for children and adolescents with aplastic anaemia or myelodysplasia. *Br. J. Haematol.* **96:** 749–756.

50. Passmore, S.J., I.M. Hann, C.A. Stiller, *et al.* 1995. Pediatric myelodysplasia: a study of 68 children and a new prognostic scoring system. *Blood* **85:** 1742–1750.

51. Kalra, R., D. Dale, M. Freedman, *et al.* 1995. Monosomy 7 and activating RAS mutations accompany malignant transformation in patients with congenital neutropenia. *Blood* **86:** 4579–4586.

52. Sokolic, R.A., W. Ferguson & H.F. Mark. 1999. Discordant detection of monosomy 7 by GTG-banding and FISH in a patient with Shwachman-Diamond syndrome without evidence of myelodysplastic syndrome or acute myelogenous leukemia. *Cancer. Genet. Cytogenet.* **115:** 106–113.

53. Cunningham, J., M. Sales, A. Pearce, *et al.* 2002. Does isochromosome 7q mandate bone marrow transplant in children with Shwachman-Diamond syndrome? *Br. J. Haematol.* **119:** 1062–1069.

54. Raj, A.B., S.J. Bertolone, M.J. Barch & J.H. Hersh. 2003. Chromosome 20q deletion and progression to monosomy 7 in a patient with Shwachman-Diamond syndrome without MDS/AML. *J. Pediatr. Hematol. Oncol.* **25:** 508–509.

55. Freedman, M.H., M.A. Bonilla, C. Fier, *et al.* 2000. Myelodysplasia syndrome and acute myeloid leukemia in patients with congenital neutropenia receiving G-CSF therapy. *Blood* **96:** 429–436.

56. Rosenberg, P.S., B.P. Alter, A.A. Bolyard, *et al.* 2006. The incidence of leukemia and mortality from sepsis in patients with severe congenital neutropenia receiving long-term G-CSF therapy. *Blood* **107:** 4628–4635.

57. Cesaro, S., R. Oneto, C. Messina, *et al.* 2005. Haematopoietic stem cell transplantation for Shwachman-Diamond disease: a study from the European Group for blood and marrow transplantation. *Br. J. Haematol.* **131:** 231–236.

58. Donadieu, J., G. Michel, E. Merlin, *et al.* 2005. Hematopoietic stem cell transplantation for Shwachman-Diamond syndrome: experience of the French neutropenia registry. *Bone. Marrow. Transplant.* **36:** 787–792.

59. Gretillat, F., N. Delepine, F. Taillard, *et al.* 1985. [Leukemic transformation of Shwachman's syndrome]. *Presse. Med.* **14:** 45.

60. Faber, J., R. Lauener, F. Wick, *et al.* 1999. Shwachman-Diamond syndrome: early bone marrow transplantation in a high risk patient and new clues to pathogenesis. *Eur. J. Pediatr.* **158:** 995–1000.

61. Strevens, M.J., J.S. Lilleyman & R.B. Williams. 1978. Shwachman's syndrome and acute lymphoblastic leukaemia. *Br. Med. J.* **2:** 18.

62. MacMaster SA & T.M. Cummings. 1993. Computed tomography and ultrasonography findings for an adult with Shwachman syndrome and pancreatic lipomatosis. *Can. Assoc. Radiol. J.* **44:** 301–303.

63. Lesesve, J.F., F. Dugue, M.J. Gregoire, *et al.* 2003. Shwachman-Diamond syndrome with late-onset neutropenia and fatal acute myeloid leukaemia without maturation: a case report. *Eur. J. Haematol.* **71:** 393–395.

64. Toiviainen-Salo, S., O. Pitkanen, M. Holmstrom, *et al.* 2008. Myocardial function in patients with Shwachman-Diamond syndrome: aspects to consider before stem cell transplantation. *Pediatr. Blood Cancer* **51:** 461–467.

65. Sauer, M., C. Zeidler, B. Meissner, *et al.* 2007. Substitution of cyclophosphamide and busulfan by fludarabine, treosulfan and melphalan in a preparative regimen for children and adolescents with Shwachman-Diamond syndrome. *Bone. Marrow. Transplant.* **39:** 143–147.

66. Bhatla, D., S.M. Davies, S. Shenoy, *et al.* 2008. Reduced-intensity conditioning is effective and safe for transplantation of patients with Shwachman-Diamond syndrome. *Bone. Marrow. Transplant.* **42:** 159–165.

67. Shah, N., H. Cambrook, J. Koglmeier, *et al.* 2010. Enteropathic histopathological features may be associated with Shwachman-Diamond syndrome. *J. Clin. Pathol.* **63:** 592–594.

68. Ho, W., C. Cheretakis, P. Durie, *et al.* 2007. Prevalence of oral diseases in Shwachman Diamond syndrome. *Spec. Care. Dentist.* **27:** 52–58.

69. Kent, A., G.H. Murphy & P. Milla. 1990. Psychological characteristics of children with Shwachman syndrome. *Arch. Dis. Child.* **65:** 1349–1352.

70. Cipolli, M., C. D'Orazio, A. Delmarco, *et al.* 1999. Shwachman's syndrome: pathomorphosis and long-term outcome. *J. Pediatr. Gastroenterol. Nutr.* **29:** 265–272.

71. Toiviainen-Salo, S., O. Makitie, M. Mannerkoski, *et al.* 2008. Shwachman-Diamond syndrome is associated with structural brain alterations on MRI. *Am. J. Med. Genet. A.* **146A:** 1558–1564.

72. Todorovic-Guid, M., O. Krajnc, V.N. Marcun, *et al.* 2006. A case of Shwachman-Diamond syndrome in a male neonate. *Acta. Paediatr.* **95:** 892–893.

73. Kamoda, T., T. Saito, H. Kinugasa, *et al.* 2005. A case of Shwachman-Diamond syndrome presenting with diabetes from early infancy. *Diabetes. Care.* **28:** 1508–1509.